THE SOLDIER IMAGE AND STATE-BUILDING
IN MODERN CHINA, 1924–1945

THE SOLDIER IMAGE AND STATE-BUILDING IN MODERN CHINA, 1924–1945

YAN XU

UNIVERSITY PRESS OF KENTUCKY

Scholarly publisher for the Commonwealth,
serving Bellarmine University, Berea College, Centre
College of Kentucky, Eastern Kentucky University,
The Filson Historical Society, Georgetown College,
Kentucky Historical Society, Kentucky State University,
Morehead State University, Murray State University,
Northern Kentucky University, Transylvania University,
University of Kentucky, University of Louisville,
and Western Kentucky University.

Editorial and Sales Offices: The University Press of Kentucky
663 South Limestone Street, Lexington, Kentucky 40508–4008
www.kentuckypress.com

Unless otherwise noted, illustrations are from "War Relief: Miscellaneous Articles and Printed Material, Undated and 1918–1940," box 95, folder 5, Records of YMCA international work in China (Y.USA.9-2-4), University of Minnesota Libraries, Kautz Family YMCA Archives. Local identifier: y_usa_9-2-4-box095-fdr005.

Library of Congress Cataloging-in-Publication Data
Names: Xu, Yan (History teacher), author.
Title: The soldier image and state-building in modern China, 1924–1945 / Yan
   Xu.
Description: Lexington, Kentucky : The University Press of Kentucky, [2019] |
   Includes bibliographical references and index.
Identifiers: LCCN 2018042195| ISBN 9780813176741 (hardcover : alk. paper) |
   ISBN 9780813176765 (pdf) | ISBN 9780813176758 (epub)
Subjects: LCSH: Soldiers—China—Public opinion—History—20th century. |
   China. Lu jun—Public opinion—History—20th century. | Military
   education—Political aspects—China—History—20th century. | Zhongguo guo
   min dang. Lu jun jun guan xue xiao—History. | Public
   opinion—China—History—20th century. | Civil-military
   relations—China—History—20th century. |
   Nationalism—China—History—20th century. | Zhongguo guo min
   dang—History. | Zhongguo gong chan dang—History. | China—Politics and
   government—1912–1949.
Classification: LCC UA837 .X785 2018 | DDC 355.10951/09041—dc23

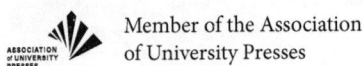

Member of the Association
of University Presses

I dedicate this book to my parents, Zhongxin Xu and Baoling Liu, my husband Sean Chao, and my daughters Evelyn, Elaine, and Elise. Without their support, understanding, and most of all love, the completion of this work would not have been possible.

Contents

Illustrations follow page 120

Author's Note

A reasonable attempt has been made to secure permission to reproduce all materials used. If there are errors or omissions, they are wholly unintentional and the publisher would be grateful to learn of them.

The pinyin Romanization system is applied to Chinese names of persons, places, and terms. This transliteration is also used for the titles of Chinese publications. Persons' names are written in the Chinese way, with the surname first, like Mao Zedong. For some popular names of people and places, well-known spellings are used, such as Chiang Kai-shek (Jiang Jieshi), Sun Yat-sen (Sun Zhongshan), and Whampoa (Huangpu).

Introduction

Politicizing the Soldier Image in Modern Chinese History

On September 3, 2015, China celebrated the seventieth anniversary of the end of World War II and its role in defeating Japan by staging an enormous military parade and celebrating a new national holiday. More than twelve thousand goose-stepping soldiers from the People's Liberation Army, together with tanks and missiles, marched in formation past Tiananmen Square and the Forbidden City in the heart of the capital. While the Chinese and Western media debated the political implication of this military spectacular, both Chinese and Western audiences were impressed by the image of uniformity and obedience that the Chinese troops presented during the parade. The Chinese state media reported that soldiers had worn through sixteen thousand pairs of shoes in the months spent practicing their goose steps. It was also reported that Chinese-built satellites guided the parade's soldiers, ensuring they did not stray more than a few centimeters from their correct spots.

Seventy years after the end of World War II, China honored her war victory. In holding such a massive military parade, it publicized to the domestic and international communities a soldier image characterized by strength and uniformity. This soldier image clearly projected Chinese president Xi Jinping's vision for the nation's future. He called it the Chinese dream: a rising power that would stand up to rivals. China was not alone in marking the end of the war by constructing an image of the soldier that promoted a political agenda. Just four months before the Chinese parade, on May 9, 2015, Russia launched a large and lavish Victory Day parade in Red Square to commemorate the Allied victory at the Eastern Front. Throughout history and in the entirety of the world's war cultures, people have made sense of, glorified, or lamented war by creating images of sol-

1

diers that range from the heroic to the pitiful; the creation of soldier imagery has said much about their societies.[1]

The scale of war in the twentieth century has proved spectacular, not only in the West but also in China. In fact, since the first major foreign conflict between China and Britain during the Opium War of 1839–1842, China has experienced wars with increasing frequency. The frequency of warfare reached its highest point during the period between 1924 and 1945. During the interwar period, when Europe struggled to recover from the devastation of the First World War, the Chinese Nationalist Party (*Guomindang*, GMD) and the Chinese Communist Party (CCP) formed the National Revolutionary Army as the First United Front to fight in the Northern Expedition (1926–1928) against regional warlords. At the end of the Northern Expedition, the GMD reunified China under the banner of the Nationalist government in Nanjing in 1927 and started to purge the Chinese Communists. During the Nanjing decade that began when the Chiang Kai-shek established the GMD government in 1927 and ended when the Second Sino-Japanese War broke out in 1937 (a period far more stable than the preceding warlord era), the GMD government launched several military campaigns to destroy the CCP army while sponsoring ambitious state-building and social engineering programs. The civil war between the two parties continued until the formation of the Second United Front in December 1936 on the eve of the full-scale Second Sino-Japanese War (1937–1945).[2] Chinese soldiers from both parties fought in the Second Sino-Japanese War, which became integrated into the Second World War after the breakout of the Pacific War in 1941.

Traditionally, war historians have examined the nature of war as it pertains to victory and defeat. The focus of such analysis is on military battles, leaders, and armies—the factors that can determine the outcome of a war. In the case of Republican China, previous scholarship on the history of war has been concerned with political policies and military strategies that could shape the process of national salvation and party struggles. This conventional approach to military history has been less concerned with either the social and cultural impacts of war or the images and experiences of the common soldiers. However, recent scholarship on military history has shifted to treat war not just as a major distinguishing characteristic of power politics but also as a cultural event—in the words of historian Andrew Huebner, "an opportunity to examine the beliefs and attitudes of human beings through their depictions of warfare."[3] Chinese military historians have started to link military history with social culture and pay attention to war experiences of individual soldiers and officers.[4] It is the intention of this book to move away from the conventional approach and add a social and cultural perspective to traditional Chinese military history. It aims to render a new way of understanding histories of war in modern China—an understanding that revolves around the construction of the discourses on the images of common soldiers at war.

The frequency of wars in China during the first half of the twentieth century was accompanied by the phenomenon that the soldier figure played an important ideological role in state rhetoric and social discussions. Different political, social, and cultural forces that shaped the history of Republican China participated in the construction of the soldier figure. They included the 1927–1949 Nationalist government of Chiang Kai-shek (1887–1975) based in the city of Nanjing, located roughly three hundred kilometers up the Yangtze River from Shanghai; the Whampoa Military Academy (*Huangpu junxiao,* China's first modern military academy),[5] commanded by Chiang; the social elites and urban publics,[6] including intellectuals, activists, professionals, writers, and students; as well as the Chinese Communists in the revolutionary base of Yan'an. The political and military authorities as well as different urban publics all employed the soldier figure as a kind of reference point to argue for their agendas and assert their political influence. The cultural negotiation of the image of the soldier thus served as a unique window to capture the dynamic interaction between the social-cultural forces and the political-military forces in Republican China in the process of state-building.

Broadly speaking, *The Soldier Image and State-Building in Modern China* tells a story about the extensive imagery of the soldier figure in the war culture of modern China. It mainly focuses on military academy cadets, lower-level army officers, and rank-and-file soldiers in regular armies—those largely neglected in most scholarship on Chinese military history or invisible in the casualty statistics.[7] This approach intends to complement, rather than side-step, the pre-existing discussion on high-level military commanders of the era, many of whom also performed critical roles as political leaders in Republican China. In terms of military branches, the discussion on images of the soldier in this book zeroes in on the foot soldiers in combat units, rather than airmen, seamen, or combat support personnel and other forces in the rear. The reason for this focus is simple: the vast majority of war-related culture in China in the first half of the twentieth century featured primarily the infantry.

The six chapters in this book are united by the main theme—the multiple meanings assigned to the soldier figure by diverse forces, and also the intentions behind those meanings. This theme serves as a useful window to explore state-building processes in the GMD and CCP areas and the complex state-society relations that were engendered by these processes in China during the first half of the twentieth century. The vast body of imagery of the soldier figure produced by different political, social, and cultural forces in modern China reveals a great deal about Chinese politics, culture, and values in this period. By examining how different political, social, and cultural forces resisted, collaborated, complicated, questioned, and confronted the heroic ideal of the soldier promoted by Chiang and the Nationalist government, this book demonstrates that the cultural negotiations over how to create and support a strong army were central to the state-building processes in modern China from 1924 to 1945, and a significant factor in determining different trajectories in state-society relations in the regions controlled by the GMD and the CCP.

From Whampoa to the Second Sino-Japanese War

This book marks the 1924 establishment of the Whampoa Military Academy as the starting point of the Nationalists' state-building process. This event symbolized the foundation of China's National Revolutionary Army (*Guomin gemingjun*), which was the military arm of the GMD state during its rule in mainland China between 1927 and 1949.[8] The book ends with China's military victory in the Second Sino-Japanese War in 1945. The construction of the soldier during the post-1945 civil wars between the CCP and the GMD remains a topic for future research. The state-building efforts and state-society relations in China between the foundation of Whampoa in 1924 and the end of the Second Sino-Japanese War in 1945, as revealed in the process of constructing the soldier figure, were greatly shaped by political, social, and cultural dynamics. These dynamics explain why creating the image of the soldier figure enlisted wide participation from diverse forces and thus became part of the state-building processes in the GMD and CCP areas. These historical dynamics also help explain why the state-society relations generated in the construction of the soldier figure presented different trajectories.

Politically and militarily, the GMD's success in the 1928 Northern Expedition military campaign allowed Chiang Kai-shek to incorporate regional warlords and their bureaucracies into the GMD state in Nanjing.

However, the political and military reunification under the GMD banner was only nominal. Although regional warlords declared their allegiance to the GMD government under Chiang, they continued to command their individual armies, over which Chiang exercised little control. As the historian Peter Zarrow comments, Chiang's authority rested on uneasy coalitions of GMD supporters, regional warlords, and local elites, and it experienced repeated challenges and rebellions due to warlord and communist rivalries.[9] As he dealt with regional warlords and the Communists, Chiang was slow in increasing his own power. Nanjing's rule was strong in only Jiangsu, Zhejiang, and Anhui in 1928; by 1931 it had been extended to Henan, Jiangxi, Hubei, and Fujian. The expulsion of the Communists from the Jiangxi hinterland in 1934 allowed Chiang to expand into southern and western China.[10] It was not until the mid-1930s that the Nationalists could claim a fair degree of control over most of China proper, "the economic heartland of the rice-producing center and the industrialized eastern cities."[11] Rivaled by the Communists and provincial warlords, Chiang and his GMD government found it necessary to legitimize their rule by strengthening ideological and cultural control. Constructing a heroic discourse of the soldier figure and extending this discourse to society was an important step in their effort.

The GMD government, which ruled China from 1927 to 1949, was above all dependent on the support of the military. After years of war, Chiang's central army eventually managed to defeat most separatist warlord armies by the late 1920s. Its success in these conflicts placed the whole of China under Chiang's centralized authority, albeit largely nominally rather than in actuality. After the Second Sino-Japanese War began in 1937, the Military Affairs Commission (*Junshi weiyuan hui*), chaired by Chiang, assumed control not only of the military but also of all administrative functions of government.[12] As the historian Lloyd Eastman argues, the Nationalist government's essential character—a dictatorial regime dependent on military force—remained unchanged throughout the Nanjing decade of 1927–1937 and the war against the Japanese.[13] For such a regime whose rule predominantly relied on military forces, military mobilization was crucial for justifying its legitimacy. The construction of the soldier figure thus became essential for the GMD government to achieve its nationwide military mobilization.

The construction of the soldier figure preoccupied not only the GMD government but also other forces in the 1930s and the 1940s because of the growing politicization of literary production and intellectuals' social activities. As literary scholar Leo Ou-fan Lee comments, by the 1930s "the artistic depth was accompanied by a sharpened consciousness of the deepening social and political crisis as the spectrum of Japanese invasion loomed large."[14] A realistic consciousness and politically motivated cultural production were manifested in several literary genres that experienced rapid development during this period, including women's autobiography and "countryside fiction" (*xiangtu xiaoshuo*). According to Jing Wang, Chinese women's autobiography during this time "favored collective action over individualistic choice."[15] No matter what the real-life situation was, the majority of Chinese women's writers "thoughtfully decided to write about their careers, their identities as writers, their roles in society, their obligations toward the country and the oppressed masses, relations with people that contribute to the building of their identity as modern, self-determined women."[16] In countryside fiction, a literary subgenre emerging in the 1930s, the writer expressed deep-seated devotion to representing the hardship, suffering, and victimization of rural residents in the face of the socioeconomic crisis and under the Japanese invasion. As Lee observes, literature featuring the countryside became "almost *ipso facto* literature of protest and dissent against a regime which did so little to ameliorate the people's livelihood."[17] Literary intellectuals voiced their political concerns by describing the life experiences of different social groups, including common soldiers.

The political consciousness of cultural intellectuals was greatly strengthened during the Second Sino-Japanese War, as many of them joined in the nationwide movement to resist Japanese aggression. They organized visiting teams and literary reporter programs, reaching out to the common masses, including ordinary soldiers on the battlefront. Their consciousness of political participation and social service was highlighted in slogans such as "Literature must join the army!" or "Propaganda first, art second!"[18] Wen-hsin Yeh points out the wartime mutual reliance between scholars working on literary and image production and propaganda units of the state. This is exemplified in her remark that "during the war, cultural production lost substantial civilian resources and support of market mechanisms, and thus had to rely on the state in various ways and at different levels."[19] The GMD government also tried to strengthen its authority and penetrate into the literary field by drawing writers into official ranks. For example, Guo Moruo (1892–1978), one of the leading writers of twentieth-century China, was appointed in 1938 as head of the Third Section of the National Military Council's newly created Political Department in charge of propaganda. Yeh's point about the mutual reliance between the literati and the state during the war echoes Leo Ou-fan Lee's comment that writers' wartime propaganda activities became "formally sanctioned by the government."[20] As this book's chapters will show, on the one hand, the nationalistic appeal embraced by the intellectuals encouraged them to cooperate with the GMD's resistance effort and to support the army; on the other hand, the intellectuals also aimed to assert their political influence in mobilizing the masses to support the army. These complex motives were tempered into the political psyche of the intellectuals, greatly impacting the construct of the soldier figure.

In the context of the nationwide resistance campaign, the literature in the early years of the Second Sino-Japanese War was characterized by "the stereotypes of guerrilla warfare and student romance and the ubiquitous note of patriotic propaganda."[21] The literary critic Hu Feng (1902–1985) pointed out the major weaknesses of war literature: (1) it gave merely neat propaganda formulas; (2) it tended to present all the trivial details without attaining any depth of vision, thus losing rather than gaining a sense of reality; and (3) in some cases it gave fantastic twists to real stories.[22] As the chief editor of the literary journal *Qiyue* (July), he encouraged battlefield reportage (*zhandi baogao wenxue*) contributors to correct these defects by adopting a personal and critical perspective.[23] This literary style had a great impact on how the reportage writers depicted images of the soldier figure.

The 1930s and 1940s were not just a period of mutual dependence between literary intellectuals and the Nationalist government; they also witnessed the increasing domination of cultural production by the Communists. For the CCP's cultural authorities, the intellectuals' literary production served politics and revolution only. The League of Left-Wing Writers, which was largely initiated by the CCP in 1930,[24] called upon league members to "pay attention to the large number of subjects from the realities of Chinese social life" and "observe and describe from the proletarian standpoint and outlook."[25] However, without consulting with important writers in the league, the CCP's propaganda chief, Zhou Yang (1908–1989), dissolved the league in 1936 and proposed "national defense literature" (*guofang wenxue*) as the official slogan for cultural production. This slogan was made to serve the CCP's 1935 Anti-Japanese National United Front (*Kangri minzu tongyi zhanxian*) policy of forcing the Nationalist government to come to some sort of coalition with the Communists and fight the Japanese. For Zhou, "the party policy of the United Front took precedence over everything else, including artistic creation."[26] The CCP continued its effort of cultural control throughout the Second Sino-Japanese War period, and its dominance reached a peak after the 1942 Literary Rectification (*Wenyi zhengfeng*) Movement launched by Mao Zedong (1893–1976).[27] The rectification, which took place at the Communist base of Yan'an in Shaanxi Province, represented a distinctive approach of the CCP to party discipline and state-building by way of "systematic remolding of human minds."[28] As the political scholar Lorena Bichler points out, the most important aspect of Mao's "Talks at the Yan'an Forum on Literature and Art" delivered in May 1942 was that it "successfully combines the proclamation of elements of an aesthetic theory in connection with persecution and criminal proceedings against dissenters."[29] The concepts and paradigms that were generated by the Literary Rectification Movement changed the life and fate of millions of Chinese people after 1949.[30] After the thought reform, intellectuals at Yan'an largely became the Communists' propaganda workers. Their construction of the soldier figure in the mass culture was intended to build the legitimacy of the CCP's social policies.

Another important development in China during the 1930s and 1940s was the emergence of urban professionals. The historian Xiaoqun Xu reveals that urban professionals organized themselves into professional associations and employed their knowledge, skills, and expertise to participate in state-building. The professional associations in Shanghai designed for various economic, social, political, intellectual, and cultural purposes "created a vibrant urban society with complex social dynamics evident in the city and beyond."[31] During the Second Sino-Japanese War, well-trained urban professionals not only actively participated in offering logistics services and recreational and educational programs to the soldiers but also advocated mobilizing the masses into that kind of work. Examples of such urban professionals were Liu Liangmo (1909–1988), a sociologist and social activist who worked at the Emergency Service to Soldiers Program (*Junren fuwubu*) of the Chinese Young Men's Christian Association (YMCA) in Shanghai; and Yu Zhaoming,[32] an expert on vocational education affiliated with the Association of Vocational Coordination for the Honorable Veterans (*Rongyu junren zhiye xiedaohui*). They wrote theoretical books on how to serve, administer, and educate soldiers, especially the wounded and disabled. They actively participated in the state-building project not only by supporting the army but also by asserting their political influence as social mobilizers and army educators. The construction of the soldier figure in their books was shaped by their social service agendas.

This book ends in the year 1945, when China emerged victorious in the Second Sino-Japanese War. The defeat of foreign enemies and the subsequent outbreak of the full-scale civil war between Chinese Nationalists and Communists intensified the tension between the GMD government and the society to a great extent. As Suzanne Pepper remarks, Chiang Kai-shek's insistence on waging a civil war against the Communists "provided the strongest possible evidence to support the charge that the Nationalist government did not exist for the people," provoking the articulation of outrage by Chinese intellectuals.[33] The student antiwar movement became one of the GMD government's most constant irritations during the civil war period of 1945–1949. The students' primary demands were an immediate end to the civil war, an end to US backing for the GMD government in that war, and a shift in public expenditure from military to civilian needs. Although student protests often aroused nationwide attention and response, the GMD government refused to accept overt opposition to its civil war policy and became increasingly ruthless in its efforts to suppress the students.[34] This book hypothesizes that the urban publics' opposition to the full-scale civil war between Chinese Nationalists and Communists made their participation in the state-building project of creating and supporting a strong army more complicated. Thus, this book leaves the post-1945 culture in regard to the construction of the soldier figure for further research.

State-Building and State-Society Relations in Modern China

Although the 1928 reunification of China under Chiang's Nationalist government remained only nominal, it still allowed the government to carry forward the undertaking of state-building. Earlier studies on the GMD's state-building in the Nanjing decade and during the Second Sino-Japanese War either emphasize the failures of the government or stress its achievements. These two strategies overlook participation in state-building by multiple social and cultural forces and the different trajectories of state-society relations generated in that process. For example, Lloyd Eastman's studies on the Nationalist government and armies argue that the inherent structural infirmities of a military-authoritarian regime lacking a base in society doomed the Nationalists to failure in their efforts to gain the loyalty of warlord troops, to tax rural society efficiently, to reform the military, and to establish a fair conscription system.[35] Julia Strauss draws different conclusions in her study of state-building in Republican China. Although the GMD's "attempt to use the Examination Yuan to build broad civil service institutions was ineffective,"[36] the workings of the Sino-Foreign Salt Inspectorate, the Ministry of Finance, and the Ministry of Foreign Affairs all challenged the conventional reputation of the GMD government as "corrupt and ineffective."[37] Strauss argues for the GMD government's capacity for "building strong and proactive institutions under exceptionally difficult circumstance,"[38] and she uses the words "surprisingly successful"[39] to describe the GMD's state-building.

Although Eastman and Strauss offer different evaluations of the GMD's state-building, they both argue that the Second Sino-Japanese War witnessed the utter debilitation and disintegration of the GMD. For Eastman, the weakening of Chiang's army during the war constituted the historical discontinuity from the prewar period. As he comments, "during the war the Nationalists' chief coercive organization, the army, ceased to be coherent and effective."[40] Strauss argues that the GMD experienced "an overall organizational paralysis and decay" during the war.[41] They are not alone in these views. Arthur Young's study of China's financial and economic development is based on the premise that, after 1937, the Sino-Japanese War and the commencement of World War II made any nation-building effort impossible.[42] More recently, David Strand has argued that the GMD's state-building projects of urban planning and railway construction were disrupted by the Second Sino-Japanese War.[43]

However, not all historians treat the Second-Japanese War as a fatal interruption to the GMD's state-building efforts. In his article on wartime political economy, Morris Bian suggests that the GMD "not only retained its capacity for institutional innovation but also greatly intensified state-building efforts."[44] Bian argues that the GMD "succeeded in creating institutions of central planning and assessment and thus increasing the rationalization of state institutions," although it failed to achieve the desired level of administrative efficiency.[45] In his book on the state enterprise system, Bian also argues that the GMD's state-building efforts were strengthened during wartime rather than weakened. In their response to the necessity of institutional change caused by the war, Bian maintains, the

GMD elite "transformed the existing mental models of institutional environments and developed new ones," which allowed them to create a new state-owned enterprise system.[46]

This study echoes Bian's thesis of the intensification of the GMD's wartime state-building efforts and contests the view advanced by Eastman, Strauss, Young, and Strand that the Second Sino-Japanese War caused a historic rupture. It makes this argument by focusing on the GMD's efforts in military training, conscription, and mobilization both before and during the war. For example, building on its successes in the previous Nanjing decade, the GMD government set up branch campuses of the Whampoa Military Academy during the war to expand its influence into regional warlords' zones. During the war the GMD kept revising the 1933 compulsory conscription law so as to strengthen its control over national armies and to militarize society. In the later years of the war, it consolidated its earlier efforts to militarize society by calling on "educated youths" (*zhishi qingnian*)[47] to join the army. These state-building agendas were designed to build a strong nation, strengthen the GMD state's legitimacy, and reinforce Chiang's personal authority.

In addition, this book provides a nuanced analysis of state-society relations in China during the 1930s and 1940s, a topic that is not covered in Bian's study. This analysis further illuminates the process and consequences of the GMD's state-building efforts. Many scholars have explored the development of local elite activism in modern China between 1924 and 1945, focusing on whether this development provides evidence of the existence of a civil society or a public sphere.[48] As Bryna Goodman suggests, the debate over the existence of a civil society or public sphere in China "leaves us with a reverse image of an all-controlling state and a society that seems stuck in tradition."[49] Her study of native place associations has shown that these associations coexisted with the increasing state penetration during the Nanjing decade. Although they were not autonomous, they still "managed to formulate a series of strategies for public activity, even to the extent of criticism of and opposition to the state."[50] Goodman thus transcends the binary opposition between state and society—a construct produced by Western historical experience.

The historian Xiaoqun Xu argues that the concept of civil society—defined as societal autonomy vis-à-vis the state—is too limiting to capture the complexity of state-society relations in Republican China. His study on Chinese professionals in the Republican state examines the concrete forms that various social groups adopted and the specific avenues that they pursued to interact with the state. Xu points out that although professional associations were allowed to exist only if they conformed to the regulations of the GMD government, they were not paralyzed by formal conformity and often managed to get around the state's efforts at control by negotiating the terms under which they would operate. Xu describes the interaction between the state and professions as interdependence. As he states, "A legitimate government could grant the professional legitimacy, while the support of the professional would in turn help legitimate the state and contribute to its modernizing projects, including modern state building."[51]

Influenced by these studies on the complexity of state-society relations in Republican China, this book does not focus on the question of whether civil society or a public sphere existed in modern China. Instead, it treats state-building as intertwined with nation-building. According to the historian Prasenjit Duara, the Chinese pattern of state-building was "closely interwoven with nation-building goals."[52] The GMD government's efforts to build modern state structures and institutions were concurrent with its advocacy of its role as the legitimate regime for the realization of the modern nation in the face of continuing domestic and foreign threats. In addition, since Chiang Kai-shek assumed control not only of the military but also of all administrative functions of government during the Second Sino-Japanese War,[53] building the nation and the state became functions of his efforts to consolidate his own authority.

This study understands state-building in modern China as a process that involved participation from multiple forces. To examine this process, it analyzes the interpretations of the soldier figure by different social and cultural forces that played important roles in modern Chinese politics, including the Nationalist government, Whampoa Military Academy cadets, regional warlords, local community leaders, urban intellectuals, activists, professionals, writers, students, and the Chinese Communists. Understanding state-building as a process intertwined with nation-building and as a process involving wide participation will help reveal multiple trajectories in state-society relations in modern China.

The GMD's heroic rhetoric of the soldier figure served the GMD's agenda of state penetration and Chiang's purpose of consolidating his authority, both of which were pursued in the name of national salvation and recovery. The GMD's state-building efforts, which included civic education (*jingshen jiaoyu*) at the Whampoa Military Academy,[54] the first compulsory conscription law (issued in 1933) and its later revisions during the Second Sino-Japanese War, the New Life Movement (*Xinshenghuo yundong*, 1934–1949), and the Campaign to Mobilize Educated Youths to Join the Army (*Zhishi qingnian congjun yundong*) in the last years of the Second Sino-Japanese War, all aimed to create a militarized, politicized, disciplined, and morally cultivated citizenry. In the GMD's political discourse, the soldier was expected to follow a code of ethics that was designed to cultivate his obedience to the political doctrine of the Three Principles of the People (*Sanmin zhuyi*),[55] his submission to discipline and regulation over personal behavior and emotional expression, and his subordination to Chiang as the leader of a hierarchical system. The GMD's political discourse of the soldier figure also constructed the soldier as a model citizen and national paragon to be emulated by the society. The soldier figure and the soldier's sacrifice in war were thus surrounded by a glorious, heroic, and romantic aura in the GMD's political discourse.

Other political and social forces layered additional meanings on the soldier figure to fit their own goals and agendas. Their interpretations of the soldier figure revealed different trajectories of their relations with the GMD government. Some Whampoa Academy cadets, regional warlords, common soldiers, and local residents had views of the soldier figure that collided with the GMD's celebration of the highly disciplined soldier ideal and its rhetoric elevating the soldier's status as a model citizen. Their views of the soldier figure revealed the tension that existed within the GMD army and between the GMD and the local society.

The relations between social elites and the state have been studied by several scholars. By showing the urban publics' relations with the state as revealed in their complex interpretations of the soldier figure, this book helps explain the ambiguous alliance between social elites and the state. Susan Glosser argues in her study on family reform in China during the first half of the twentieth century that the rearrangement of family organization and gender roles in the Republican era resulted in an alliance between social reformers and the state. Glosser argues that "reformers and revolutionaries of all political stripes welcomed state intervention and willingly subordinated individual rights to the demands of the state in return for its promise to save China."[56] The alliance between social reformers and the state ultimately facilitated state dominance of society and impeded the potential for popular challenges to state authority. Feminists, intellectuals, politicians, and entrepreneurs were all unable to counter the state's demands.

Glosser observes that the welcoming of state intervention among elite publics characterized the political culture throughout the Republican period. However, this observation is questioned in Eugenia Lean's work. In her study of the sentiment-centered and sensation-creating case of the 1935 killing of a warlord by the woman Shi Jianqiao, Lean employs as her theme the role of public passions in creating urban publics and in demarcating the boundaries among ethical, judicial, and political power in the building of modern China.[57] Lean demonstrates that the sensationalism of the mass media in the 1930s might have helped to mobilize, or hail into being, multiple urban publics that expressed a powerful critique of an actively centralizing state, although this force remained vulnerable to manipulation by higher authorities. Instead of welcoming state intervention, intellectual commentators articulated critical views and asserted their own political agendas and institutional influence. In other words, for Lean, social reformers were not in an alliance with the state; they were critical of state intervention.

This book argues that the various urban publics, including intellectuals, activists, professionals, writers, and students, both allied with the state (intentionally or not) and were critical of state intervention. The urban publics' relations with the state, as revealed in their responses to the GMD's heroic discourse of the soldier figure, were more ambiguous. After the Second Sino-Japanese War broke out, urban intellectuals, activists, and professionals actively participated in the war service, reporting the war, supporting the soldiers, and writing books to publicize their experiences in performing army service. Literary writers wrote fiction and battlefield reportage to reveal the brutality of the war and search for solutions for the nation's salvation. In 1944, when the Nationalist government launched the Campaign to Mobilize Educated Youth to Join the Army to address both the international situation and the escalating conscription crisis at home, many college students answered Chiang Kai-shek's call by joining the Youth Army (*Qingnianjun*).[58] These urban publics' activities supported the national resistance against Japan and campaigns launched by the GMD serving the soldiers and reforming the army. Their participation in supporting the army during the national salvation movement represented their efforts in state-building.

Although the urban publics did not resist the state's efforts to militarize society, they still managed to maintain some level of independence by complicating or de-idealizing the political discourse of the soldier figure in many ways. The intellectuals and professionals echoed the GMD's discourse about the soldier in the sense that they highlighted and praised the soldiers' bravery in their war reports. However, the intellectuals and professionals challenged the state discourse by describing the soldiers as vulnerable, poorly educated, eager for emotional support and vocational training, motivated to fight by personal pursuits, or active in practicing self-government. In doing so, they criticized the GMD's organization of soldier relief work and what they viewed as endemic corruption within the GMD army. Additionally they asserted their political influence as social mobilizers, social critics, army educators, and promoters of democracy.

Treating state-building as a process involving participation from multiple forces will also show its gendered aspect. For instance, this book demonstrates that the political culture surrounding the construction of the soldier figure as well as the impact of war on people were highly gendered. Many scholars have adopted gender as a category of analysis in their studies on state-building and state-society relations in Republican China. For example, Lean demonstrates that assassin Shi Jianqiao's gender was crucial in engendering public sympathy that generated the new conceptualizations of publics. Shi's filial heroism "effectively enacted the moral virtues of the new Chinese nation," and her ardent broadcast through the mass media "mobilized, as well as symbolized, the collective sympathies of the authority of an emerging public."[59] At the same time, left-leaning writers and judicial reformists, who became uncomfortable with the threat posed to their rationality by the rising mass public, "denigrated the collective emotionalism as foolhardy and feminine."[60]

This book argues that the political culture related to the construction of the soldier image and the debates about how to build and support the army in the Republican era were gendered in several ways. The GMD state claimed in its first compulsory conscription law, issued in 1933, that military service was the duty only of male citizens; in the middle of the Second Sino-Japanese War it stipulated that women, too, had military duty, but that they were only allowed to perform noncombat tasks. Although female intellectuals and activists in the GMD areas did not challenge the GMD's stipulation, they asserted their political influence by describing the soldiers as vulnerable and affectionate persons who needed care from the larger social masses, even including children. The writer Xie Bingying (1906–2000), who claimed herself to be a woman soldier, did not depict the soldier as vulnerable to war brutality, as the male writers Xiao Jun

(1907–1988) and Qiu Dongping (1910–1941) did. Instead, she portrayed herself as a brave rebel who opposed traditional gender roles and struggled for personal independence. In the CCP's *yangge* movement, the way that peasants showed their concern for the soldiers was also gendered. In the *yangge* dramas celebrating army-people solidarity advocated by the CCP, the female peasants provided support to the soldiers by performing housework, such as cooking. Gender serves as a useful analytical category in the context of creating soldier images because it shapes the political culture with respect to military service in GMD areas and the mass culture in regard to army-people relations in CCP areas.

War and Soldiers in Modern Chinese Military History and Literature

This book aims not only to shed light on state-building and state-society relations in the period between 1924 and 1945. It also attempts to open up a new avenue to the discussion of (or the studies on) the war's impact on the society and popular conceptions of the soldier during the war. The existing literature has tended to neglect or underestimate the role of war in modern China. As David Graff and Robin Higham point out, English-language literature on Chinese military history, ancient or modern, was extremely limited two decades ago.[61] Previous studies on soldiers and wars in modern China focus mainly on officer training, army command, warlord politics, military actions, armaments, the careers of individual warlords, and the role of foreign advisers.[62] Chinese military history has developed into an important subfield of Chinese history during the past decade, but most existing studies are written mainly from political and military perspectives. The popular perceptions of the soldier and the construction of the soldier figure by social and cultural forces, such as intellectuals, professionals, and young students, remain largely unexamined.

Few books are devoted to the study of ordinary soldiers in modern China. Scholarship on Chinese armies and soldiers is mainly dominated by biographies of leading warlords or studies of major warlord factions and their politics.[63] China historians have tended to view the Warlord Period (1916–1928) as representing "the worst aspects of China: lack of concern for the nation, lawlessness, social disorder, corruption, factionalism and lack of any moral concerns."[64] With this view of warlordism, historians see soldiers as victims of the dark sides of warlordism and as representatives of army corruption and social disorder. Diana Lary's study of Chinese soldiers in warlord armies was the first to focus exclusively on ordinary soldiers between 1911 and 1937. Lary analyzes the backgrounds, experiences, and lives of soldiers. She argues that the predatory behavior of soldiers grew out of the brutal conditions they faced in the service of their warlord.[65] Scholars on Chiang Kai-shek's Nationalist state mostly treat soldiers as faceless numbers in the Nationalist army who either fought bravely during the Second Sino-Japanese War or were the "seeds of destruction" of Nationalist China.[66] However, these studies do not examine the ideological and cultural construction of the soldier by the government and other social and cultural forces.

One of the accomplishments of Chinese military history studies in the past two decades is the detailed examination of the effects of military affairs and wars not only on politics but also on broader aspects in modern Chinese society, such as industry, culture, and social mentalities. For example, in his study of warfare in 1924 between the Fengtian and Zhili warlord factions, Arthur Waldron argues that the arms race between them drove China's industrialization and commercialization.[67] Eugene Levich presents the Guangxi warlords during the 1930s as neglected soldier-statesmen and veritable agents of reform who achieved significant accomplishments in mass education, mobilization of women, road construction, military recruitment, and militia training.[68] This book follows this recent scholarly trend by paying attention to the impact of the war on common people, including soldiers. It approaches the impact of the war on multiple social and cultural forces from a cultural-political perspective. It is the only study so far that focuses on the multiple meanings of the Chinese soldier figure created by different social and cultural forces.

To achieve the analytical goal of using military history to illuminate the processes of state-building and explore the dynamics in state-society relations during the period, this book makes extensive use of literary sources on the soldier image to analyze the relations between the intellectuals and the state in modern China. The soldiers in fiction and battlefield reportage were not faceless numbers in history studies. They had personal emotions and needs, they talked and thought, and they experienced psychological transformations before, during, and after combat. The soldiers' experiences of fighting the war and serving in the Nationalist army, as depicted in literary works, reveal a great deal concerning the writers' thinking about the national crisis and solutions and about their roles in the relations with the state. Timothy Brook suggests that, for soldiers, "wartime suffering means combat injury and death in the first instance but also fear and anxiety, brutalization and dehumanization, and extraction from the routines that constitute normal life."[69] Literary writings treat soldiers as ordinary human beings and thus provide a vivid description of their fear and anxiety, and therefore they are useful in studying the impact of the war on common people, especially the soldiers. The literary creation of soldiers will also reveal the writers' perception of how to build and support the armies, their attitudes toward phenomena in the society and in the army, and the roles the writers identified for themselves. Therefore, literary sources can show both how intellectuals participated in state-building and what their relations with the government were.

Although adding a literary perspective to the conventional writing of Chinese military history is a strength of this book, my purpose here is not to provide a comprehensive overview of all possible representations of soldiers in Chinese war literature. Instead, I intend to show the wide range of soldier images created by various social and cultural forces with different goals. In particular, this book provides a close examination of three writers—Xiao Jun (1907–1988), Qiu Dongping (1910–1941), and Xie Bingying (1906–2000). The reasons for this are threefold. First, their backgrounds are representative of literary workers, male and female, who had direct army or even combat experience. Second, their works represent the literary genres that were most popular in wartime China—fiction, battlefield reportages, and autobiographies. Third, these three writers created diverse images of soldiers in their works that varied extensively, from CCP-led guerrilla soldiers, to mid-level Nationalist officers, to women soldiers. A close reading of their works shows not only the complexities of the cultural discourses of soldier images in Chinese war culture, but also the gendered aspect of wartime Chinese society.

The process of state-building involved multiple forces. This book makes particular efforts to explore these forces by examining the activities and writings of many historical actors that were not touched by previous studies. Their thoughts about how to support and serve the soldiers shed light on the relations between urban professionals and the government. For example, the social activist Liu Liangmo and the vocational education expert Yu Zhaoming, as well as the associations they had affiliations with, the Emergency Service to Soldiers Program of the Chinese YMCA at Shanghai and the Association of Vocational Coordination for the Honorable Veterans, have not been previously discussed in English-language studies on the Second Sino-Japanese War. When the soldier images described in Liu's and Yu's writings are read with an understanding of the heroic discourse of the soldier figure deliberately constructed by Chiang and the GMD state, the relations of these professionals with the state become more clear. Other important examples are the writings of educated youths in the Youth Army, which have been less examined in Chinese military studies. The students cooperated with the GMD government's state-building efforts by joining the army upon Chiang's call in 1944. However, the student soldiers' experiences as described in their writings reveal that they tried to forge a new discourse on the soldier image by stressing the principles of self-government and democracy in building the army.

Social Emotion and Mass Mobilization in the Chinese Communist Revolution

This book will also contribute to the scholarship on the Chinese Communist revolution by analyzing the role of social emotion in state-building. Existing studies examine the CCP's wartime success in military mobilization using a paradigm of double appeals—national resistance and socioeconomic and political reforms. This framework is based on the presumption of rationality in decision-making and behaviors on the part of the CCP and on the part of the peasants.[70] The political role of emotion has recently attracted the attention of scholars on state-building and the Chinese Communist revolution. Experts on the May Fourth movement and Communism discuss how various emotions that were directed toward different things accompanied and affected radical intellectuals' political action. Some examples would be love for China, hatred for the status quo, anger with exploitation, enthusiasm in the revolutionary cause, and passion for political ideology.[71] As the historian Hung-Yok Ip comments, the choice of radical politics is always imbued with emotion.[72] Ip's study on Chinese intellectuals defines emotion as "a variety of feelings that accompany the individual's intellectual activities, behaviors and actions."[73] Ip shows that Communist intellectuals embraced the ideological imperative of antielitism by humiliating themselves as being in need of advice from the masses. But ironically, intellectuals also benefited from this self-humiliation in that they ennobled themselves by expressing their willingness to sacrifice for and be loyal to the revolution. Ip suggests that the CCP made use of revolutionary intellectuals' self-construction as heroic and sophisticated figures to bolster its legitimacy.

Ip's study mainly examines the self-construction of the Communist intellectuals. How the CCP mobilized emotion for its political goals still remains largely unaddressed. The historian Elizabeth Perry is one of the few scholars who brings the CCP's mobilization of emotion into focus. Her article "Moving the Masses: Emotion Work in the Chinese Revolution" draws attention to the largely neglected feature of the revolutionary process: the mass mobilization of emotions. Perry argues the Communists systematized emotion work as part of a conscious strategy of psychological engineering, and that they accomplished this by building on preexisting traditions of popular protest and political culture.[74] Chinese scholar Yu Liu picked up this theme in a recent article, "Maoist Discourse and the Mobilization of Emotions in Revolutionary China." Liu examines how Maoist discourse engineered revolutionary emotions as a method of political mobilization. Liu identifies three themes of the Maoist discourse, each of which aimed at provoking one type of emotion: the theme of victimization, which mobilized indignation in struggle campaigns; the theme of redemption, which generated guilt in thought reform campaigns; and the theme of emancipation, which raised euphoria in social transformation campaigns. Liu also points out the techniques of producing these emotions, which were propagation, personalization, magnification, and moralization.[75]

Lean's study of the political, cultural, and legislative developments surrounding the Shi Jianqiao assassination case also analyzes the function of popular sympathy in state-building, although her study deals with the GMD state. Lean demonstrates that overwhelming public sympathy toward the assassin Shi, which was engendered in and empowered by the mass media, mobilized into a force of multiple urban publics that threatened the moral authority of the cultural elite and influenced the state's tactics in legitimating its power.[76] She suggests that multiple urban publics who publicized their passion in urban media could potentially function as civil society in 1930s China by forcing the Nationalist government to respond to popular sympathy. In other words, public emotion that was engendered in urban media empowered multiple publics. Emotion was defined in her research first as sympathy that urban publics felt toward the woman who assassinated a warlord, and second as filial piety, individual-based social emotion between a son/daughter and a parent, which was a politically, socially, and culturally sanctioned morality and emotion.

Stimulated by these studies, my chapter on the construction of the soldier figure in mass culture in wartime Yan'an examines the emotional dimensions of mass mobilization as a key ingredient in the CCP's success in building the state and forging social integration at its revolutionary base. By defining the soldier figure along the line of the emotional bond with the peasants, the CCP tried to build the bond between soldiers and peasants as a legitimate category of social relations and emotion. The CCP tried to build the group-based emotions between soldiers and peasants as a politically and culturally sanctioned social emotion. The CCP's strategy of state-building by mobilizing and engineering social emotions distinguished it from the GMD's approach of elevating the soldier's status as model citizens to be emulated by the public. The forging of this social emotion was an indispensable strategy for the CCP's effort to justify its legitimacy and manage closer social integration in its wartime revolutionary base.

Organization and Sources

This book is organized into six thematic chapters. This organizational structure is based on the theme of the research—multiple meanings of the soldier figure assigned by diverse forces, as well as the intentions behind these meanings. This structure will help reinforce my thesis that the soldier figure served as a medium of articulation in which different political, social, and cultural forces in modern China argued for their political influence and asserted their visions of state-building and state-society relations.

The first two chapters examine the heroic discourse of the soldier figure constructed in Chiang Kai-shek's speeches on civic education at Whampoa (chapter 1) in the first few years after the 1924 establishment of the academy, and in the GMD's military laws and political propaganda in the 1930s and 1940s (chapter 2). These efforts by the GMD at state-building all aimed to train the soldier into becoming the model of a politicized, disciplined, militarized, and morally cultivated citizen who was obedient to Chiang's leadership and submissive to the Three Principles of the People. Whampoa cadets who cultivated a code of ethics were constructed as model soldiers to be emulated by national armies. The GMD's political discourse of the soldier figure glorified the heroic aspects surrounding him, such as bravery, military discipline, emotional suppression, loyalty to orders and leaders (especially Chiang), and sacrifice in combat. These two chapters conclude with a brief discussion of the meanings that Whampoa cadets, regional warlords, common soldiers, and local *baojia* leaders gave to the soldier ideal advocated by Chiang and the GMD government. The different meanings given by these forces reveal the tensions that existed within the GMD army and between the GMD state and local society.

The discussion of the GMD's state-building efforts draws primarily from Chiang's speeches and lectures delivered to Whampoa cadets, the conscription laws, military edicts, documents by GMD political and military authorities, literacy textbooks for normal soldiers, and New Life Movement propaganda works. These sources help explore the political discourse of the soldier figure, the intentions of Chiang and the Nationalist government in forging this discourse, and the techniques they employed in justifying this discourse. The memoirs of Whampoa graduates analyzed in chapter 1 not only confirm Chiang's intentions and techniques of politicizing and disciplining cadets; they also reveal that Chiang's state-building goal of extending the soldier ideal promoted at Whampoa to the army and society met resistance from Whampoa cadets, civilian society, and regional warlords.

Chinese secondary scholarship on soldiers' treatment, memoirs of Nationalist authorities, archives of the Nationalist government, and English secondary scholarship on the actual implementation of the compulsory conscription system are analyzed in chapter 2. These sources show that despite the GMD's efforts at elevating the soldier's status as model citizen and epitome of morality in the 1930s and the 1940s, the social masses still associated being a soldier with misery and did not embrace the heroic aura surrounding the soldier figure. The GMD's state-building effort of producing politicized, disciplined, militarized, and morally cultivated citizenry by constructing the soldier as the model citizen and epitome of morality was resisted by the common masses and regional warlords and even confronted by some Whampoa cadets.

Chapters 3, 4, and 5 will examine the construction of the soldier figure by different urban publics, including intellectuals and professionals, writers with direct army experiences, and young students during the Japanese occupation of the northeastern parts of China in the 1930s and during the Second Sino-Japanese War between 1937 and 1945. They all actively participated in state-building and asserted their political influence in their participation. The ways they constructed the soldier figure were shaped by their agendas and revealed the ambiguous alliance between them and the GMD government. Chapter 3 shows that after the breakout of the Second Sino-Japanese War in 1937, urban intellectuals and professionals supported the army and succored wounded and disabled soldiers. In their participation, they asserted their role as social mobilizers and army educators. They showed their respect for the soldiers' bravery, but they also complicated the GMD's heroic soldier ideal by stressing soldiers' suffering, emotional needs, and poor education. Chapter 4 discusses Xiao Jun's 1935 fictional *Bayuede xiangcun* (Village in August) and Qiu Dongping's battlefield reportage written in the first two years of the Second Sino-Japanese War. It shows how the writers Xiao Jun and Qiu Dongping performed their roles as social critics by revealing the dark sides of the society and the army and envisioning the basis for national salvation and the creation of a strong army. The construction of the soldier figure was also gendered in the literary writings. Although Xiao and Qiu questioned the GMD's heroic soldier ideal and highlighted the vulnerability of soldiers, the woman writer Xie Bingying portrayed a brave and strong image of herself as a woman soldier. Stressing her bravery fit Xie's goals of breaking the confinement of traditional gender roles and struggling for independence, as well as asserting women's political influence. Chapter 5 shows that educated youths, especially college and high school students, actively participated in state-building by joining the army in the later years of the Second Sino-Japanese War. They advocated the importance of practicing self-government and democracy as the way of building the army. Newspaper reports and literary writings in the genres of autobiography, fiction, and reportage as well as the writings by the student soldiers constitute the largest body of primary sources in these three chapters.

Chapter 6 shows that during the Second Sino-Japanese War the CCP employed the state-building strategy of transforming the attitudes and emotions of the soldiers and intellectuals toward the peasants. They constructed the soldier figure within the framework of army-people solidarity, and the CCP's cultural workers celebrated the emotional bond between soldiers and peasants in *yangge* dramas.[77] The *yangge* is treated as a representative genre of Communist mass culture because it was officially developed into a mass movement in 1943. Chapter 6 is based on the following Chinese and English primary sources: party policies on how to direct and facilitate the cultural activities in Yan'an, party and army leaders' conference talks and work reports that summarize the popularity of cultural activities among soldiers and peasants, biographies and memoirs of political, military, and cultural figures who lived in Yan'an during the war, selections of literary and artistic productions, and Western observers' firsthand accounts of the social lives in Yan'an. The insiders' and outsiders' views of the social life and popular culture in Yan'an will work together to decode the construction of the soldier figure. The *yangge* dramas that are closely examined in this chapter are drawn from the volume of the *Yan'an wenyi congshu* (Series of Yan'an Literatures and Arts) on the *yangge*. This volume contains the largest selection of *yangge* plays so far available. The twenty-eight *yangge* plays selected for the volume are those that were frequently performed and enjoyed wide popularity in wartime Yan'an.

The images of the soldier figure examined in my book mainly come from the conscription laws, political propaganda, literacy textbooks for soldiers, Chiang's lectures and speeches, guidance and theoretical works on soldier service written by urban professionals, memoirs, autobiographies, fictions and reportages written by army-affiliated writers, and the *yangge* dramas in Yan'an. Except for the military edicts, political propaganda, and Chiang's speeches, this book selects a few cultural genres and a few well-known literary texts to demonstrate the ubiquity of the meanings of the soldier figure to urban publics; it does not aim at a comprehensive coverage of the soldier figure created in any cultural genre in modern China. The book aims for a deeper understanding of the complex images of the soldier figure by reading these diverse primary sources against each other. This technique supports the analytical goal of exploring the state-building intentions of different forces in constructing soldier images.

Although the different genres of historical and literary sources examined in this book are useful in revealing the political culture around the construction of the soldier figure, some of the genres should be treated carefully. One such example is the memoirs of Whampoa graduates, which were written several decades after their graduation from Whampoa and selectively published in 2010. *The Soldier Image and State-Building in Modern China* employs these memoirs not as the backbone of the sources to examine the construction of the soldier images at the Whampoa Military Academy. Instead, it uses these memoirs mainly to confirm soldier ideals advocated by Chiang Kai-shek.

Another type of source that needs to be read carefully is the autobiography, a historically situated practice of self-representation. Chapter 4 of this book selects Xie Bingying's autobiographies, such as *Yige nübingde zizhuan* (the 1936 autobiography of a female soldier), to discuss the construction of the soldier figure by a female writer and soldier. Jing Wang has remarked that autobiography was "dismissed as mere personal accounts unworthy of critical attention" and that it has been neglected as an independent category in Chinese literary criticism.[78] Nonetheless, this book contends that Xie's autobiography is helpful in revealing the construction of the soldier figure because it "places the writer at the center and requires the readers to see her life and her person the way that she designs them."[79] This book does not aim to seek historical facts about Xie's army life from her autobiography, but instead tries to explore her personal and subjective world as a female soldier.[80]

Since this book chooses to stress thematic patterns across genres, it does not aim to thoroughly historicize the transformation of construction of the soldier figure through different periods. The framework that guides this book is that resulting from the multiple meanings of the soldier figure in modern China rather than the historical transformation of the soldier figure. This framework attempts to demonstrate that the discourses of the soldier figure constructed by social and cultural forces coexisted but competed with the GMD's political discourse.

The multiple meanings of the soldier figure constructed by the GMD and CCP governments in Nanjing and Yan'an as well as different groups of political, social, and cultural forces, which coexisted and competed with each other, serve as the lens through which a deeper and more nuanced understanding of various trajectories of state-society relations in Republican China can be gained. The GMD's state penetration compromised with instead of eliminating the growing political participation of multiple forces; at the same time, the CCP managed wider social mobilization and closer social integration by remolding the emotions of intellectuals, soldiers, and peasants.

1

Training Model Soldiers at the Whampoa Military Academy

The story of politicizing the soldier image in modern Chinese history starts at the Whampoa Military Academy (*Huangpu junxiao*),[1] the "West Point of the East." The academy was founded by the Chinese Nationalist Guomindang Party (GMD) in 1924 with the assistance of the Soviet Union and was led by Chiang Kai-shek as the commandant and Liao Zhongkai (1877–1925) as the party representative. It was established as the foundation of the National Revolutionary Army, which later became the military arm of the GMD state during its rule in mainland China between 1927 and 1949.

The Whampoa Academy serves as an ideal case study to begin the analysis on the construction of the soldier figure in modern Chinese state-building because its intended special significance by the GMD leadership resulted in its having a substantial impact on Chinese military modernization. Whampoa produced many important commanders who fought in many of China's conflicts in the first half of the twentieth century, notably the Northern Expedition (1926–1928), Nationalist-Communist civil wars (1927–1937), and the Second Sino-Japanese War (1937–1945). The academy was not simply a military institution with a singular mission to train military cadets who could become model soldiers and competent Nationalist army officers. It also had an inherently political color in its nature. Chiang, as the commandant of the academy, valued it as a political tool that he could use not only to win the hearts and minds of national military officers but also to centralize China's local military forces under his own command.

The inauguration of the academy took place offshore from the Whampoa docks, some fifteen miles downriver from Guangzhou, a port city northwest of Hong Kong on the Pearl River. In the same year, the Training Regiment of Whampoa Military Academy (*Huangpu junxiao jiaodao-*

tuan), also called the Academy Army (*xiaojun*), was founded. In 1925, the GMD announced the organization of the National Revolutionary Army, with Whampoa's training regiments joining it to form its first division.[2] In response to the April 12 Incident of 1927, when Chiang and conservative factions within the GMD violently suppressed the Chinese Communist Party and other leftwing organizations, in 1928 the Whampoa Military Academy was renamed the Central Army Officer Academy (*Zhongyang lujun junguan xuexiao*). The academy relocated to the eastern city of Nanjing, the capital of Nationalist China, in 1929 and then to the city of Chengdu in the southwestern interior after the Second Sino-Japanese War broke out in July 1937. It returned to Nanjing in 1946, and three years later moved to Taiwan along with the Nationalist government.[3] Between 1924 and 1949, the academy recruited twenty-three regular classes, set up over ten branch campuses, launched numerous short-term training programs for contemporary army officers, and cultivated over forty thousand military elites for the Nationalist army.[4]

This chapter examines Chiang's speeches and lectures to Whampoa cadets in the early years after the establishment of the academy, as well as Chiang's efforts to build branch campuses during the Second Sino-Japanese War. This framework emphasizes the thesis that the GMD strengthened its state-building efforts after the breakout of the war. It also helps explain the multiple functions of Whampoa for Chiang and regional warlords. The memoirs of Whampoa graduates concerning their lives and training at Whampoa, which were written in the 1960s, confirm Chiang's intentions behind his commandership and reveal complex tensions between Chiang and regional warlords and among Whampoa cadets.

Regulating the Mind

From the late Qing dynasty (1644–1911) onward, China suffered repeated humiliating defeats in the Opium Wars and other foreign conflicts. Among China's intellectual elites, the dominating perceptions on how to construct a state or conduct revolution viewed the military as "the midwife of a modern and cohesive China."[5] However, as the historian Colin Green comments, the warlord era in the late 1910s and early 1920s had "shattered the Chinese people's faith in the military as a positive institution," and soldiers were regarded "as little better than bandits."[6] Restoring the tarnished image of the military and of soldiers was considered an utmost priority by the Chinese Nationalists who sought national unity and revival. In 1921, after his talks with the representative of the Comintern, Henk Sneevliet (1883–1942), Sun Yat-sen (1866–1925)—the first president and founding father of the Republic of China—realized that it was not the ignorance of common soldiers but the lack of nationalist awareness and political beliefs among military officers that doomed any effort toward a real revolutionary army to failure.[7] Sun believed that the building of a party army should begin with training model soldiers and potential cadres for the revolutionary force.[8] Under the policy of alliance with the Soviet Union and the CCP (*lian E rong Gong*), Sun was influenced by the experiences of the Red Army of the Soviet Union and intended to build an army subject to strict political control by the party. In 1924, Sun decided to found the Whampoa Military Academy, seeking to establish a reliable and well-trained armed force to support his nationalist revolution.

Sun sent Chiang Kai-shek to Moscow to conduct preliminary research on the Soviet military system for four months in 1924, and Chiang became Whampoa's first commandant. He remained in this position until 1947. Whampoa's military curriculum was "set up under the guidance of the Soviet advisory group, utilizing the latest military theories and techniques, albeit with a distinct Soviet flavor."[9] What made Whampoa fundamentally different from other military schools was that from the very beginning it accepted the Russian model of putting political control at the center by emphasizing political training and ideological education in order to train an officer corps loyal to the GMD. Topics covered in Whampoa's political curriculum mainly included the Three Principles of the People and China's revolutionary history. Because of Whampoa's significant role in the development of the GMD's military forces, I suggest a reperiodization of the GMD government founded in 1927 by treating the 1924 establishment of Whampoa as one of its earliest state-building efforts.

In 1924, Sun Yat-sen elaborated the mission of Whampoa and his expectations of the cadets in the "Diyiqi zhaosheng jianzhang" (General regulations on the first class enrollment): "The revolutionary army consists of soldiers who are dedicated to saving the nation and its people. The cadets in the Whampoa Military Academy would be the backbone of this army. . . . So our students should not be afraid of death; instead, they should follow the path of revolutionary pioneers. Our first class of five hundred cadets is expected to be the foundation for an ideal revolutionary army. With this ideal army, our revolution can succeed and China can be saved."[10] His comments on the mission of the cadets revealed that when the academy was founded in 1924, it was to be devoted to training model soldiers for the national military forces to achieve national unification and salvation. This principle was also revealed in the first item in the "Lujun junguan xuexiao kaoxuan xuesheng jianzhang" (Memorandum on testing and selecting cadets for the army officer school) issued by the GMD on March 2, 1924: "Aiming to improve the national army, our academy provides ambitious and enthusiastic young people in our nation an opportunity to study military arts. The teaching of the Three Principles of the People at our academy will cultivate strong political beliefs among the cadets and enable them to assume the offices of lower-level cadres in the army."[11]

The registration materials cited here show that Whampoa—the institution established by the GMD to build a unified and powerful nation—aimed from the very beginning to produce well-trained, dedicated, and politically indoctrinated soldiers and to extend these ideals to the national army and citizens. This point is reflected in Green's argument that, "far from training the cadets to be nothing more than obedient servants of the Guomindang, Whampoa was from the beginning intent on preparing them to assume what it believed to be the soldier's proper role in a modern society—that of champion and moral exemplar for all citizens in the struggle to build a rich and powerful nation."[12] The academy was built to serve, in effect, as a laboratory or a crucible for casting a new discourse on the citizen ideal that stressed political indoctrination, military discipline, and moral cultivation.

With the goal set to train model soldiers and potential cadres for the Nationalist army, the academy implemented the strategy of including civic education in the military training. In Green's words, "Chiang Kai-shek set out to rehabilitate the military profession by means of a spiritual training program that emphasized a strict code of ethics."[13] The emphasis on civic education differentiated Whampoa from the curriculums of earlier military academies.[14] The sense of ethics "distinguished the revolutionary soldiers from the warlord rabble in the eyes of a skeptical public and restored popular trust in the military."[15]

The importance of civic education was repeatedly emphasized in the speeches and lectures by political authorities. Luo Derong (1903–1947), a graduate of the academy's third class who became a major general in the Nationalist army in 1945, wrote in the preface to his book *Xinbian junren jingshen jiaoyu* (New edition of civic education for soldiers) in 1932, "Do you deserve the title of a soldier only by wearing a military uniform and carrying a weapon? Without a military spirit, you cannot be considered a soldier, especially a revolutionary soldier."[16] Civic education was considered a training requirement in order to convert the previously undisciplined cadets into highly politicized and well-disciplined soldiers who could serve as models for the national army.

As the commandant of Whampoa, Chiang Kai-shek exerted strong influence over the civic education by constantly lecturing and admonishing the academy cadets. Chiang justified the necessity of civic education by appropriating the concepts in traditional military works. He reinterpreted the principle of the five components in warfare proposed in *Sunzi bingfa* (Master Sun's *Art of War*), written by Sun Tzu during the Warring States Period (403–221 BCE), to assert the significance of political discipline and ideological indoctrination. The five components of war in *Sunzi bingfa* were *dao* (moral law, the masses' absolute trust in the ruler), *tian* (heaven, good timing), *di* (earth, geographical advantages), *jiang* (the commander, key leadership qualities of wisdom, sincerity, benevolence, courage, and strictness), and *fa* (method and discipline, organization and management of personnel and supplies).[17] Chiang Kai-shek equated the art of *dao* with the Three Principles of the People.[18] In Chiang's eyes, sharing the same political belief ensured the absolute submission of the soldiers to his leadership within the academy, and by extension the army. He strongly held to the notion that a model soldier for the national army should be reformed by arming his mind with the doctrine of the Three Principles of the People.

The memoirs of Whampoa graduates show that the academy stressed the importance of ideological indoctrination and exerted its ideological control over the cadets by regulating their personal lives. Shen Zhenchuan,[19] a graduate of the academy's eighth class (1930–1933), recalled in July 1964 that the academy exerted strict control over students' speech and behavior, checking students' letters and diaries, examining students' extracurricular readings, and eavesdropping on students' discussions. After entry into the school, there were screening tests in the first three months and again after the first year. The standards for the tests were not military learning or physique; rather, they assessed whether there were any discernible associations with Communists and any mentality sympathetic or supportive of the Communist movement. Cadets who contacted relatives and friends in the Communist areas, who had correspondence with progressives, who showed discontent with the Nationalist officers, even those who covered their books with red paper, were all likely to fail the screening tests and receive punishments, such as dismissal (*tuixue*), repeating the year's work (*liuji*), writing letters of confession (*huiguoshu*), and physical confinement (*jinbi*).[20] The Whampoa graduate's memoir confirms that strong political belief in the Three Principles of the People was a key characteristic in Chiang's citizenship ideal, and that was made clear in the model soldiers' training at Whampoa.

Disciplining Personal Behavior and Emotional Expression

Civic education at Whampoa tried not only to cultivate a strong political belief in the minds of the cadets but also to dominate the cadets' understandings of their personal behavior[21] and physical sacrifice. In the speech "Keku nailao yu kangkai xisheng zhi biyao" (Necessity of diligence, endurance, and heroic sacrifice) delivered to Whampoa cadets in 1924, Chiang claimed, "After you enter the military academy, your bodies belong to the party state, and so do your lives."[22] The rules of Whampoa and the speeches by Chiang show that he tried to exert control over the cadets' personal behavior in their daily routines.

The Whampoa Academy recruited five hundred to seven hundred cadets for each class, drawing its student pool from a variety of sources that included students, workers, and peasants, all of whom had widely differing educational backgrounds. According to the "Huangpu junxiao diyiqi xueyuan ruxue beijing qingkuang yilanbiao" (Survey table of the background of the cadets enrolled in the first class of the Whampoa Military Academy),[23] the cadets pursued a variety of professions before they entered Whampoa. Many of the cadets had prior work experience. For example, 39 of the 135 cadets in one of the six units in the first class had received military education, or had served as soldiers or lower-level officers in local warlord armies or militias. Others had extremely diverse civilian educations, such as law, drama, art, and foreign language. Still others had equally varied previous professions, including sports instructors, elementary school teachers, editors, and civil servants.[24]

To discipline these cadets with such diverse work experiences and educational backgrounds, between 1924 and 1927 Whampoa issued a series of statutes to regulate their personal behavior. Examples of such statutes were the "Gemingjun xingshi tiaoli" (Penal clause in the revolutionary army), the "Xiao zhixingguan qinwu guize" (Regulation on the logistics of the duty officer), the "Chishou lijie ling" (Order of following military rituals), and the "Chongshen jingli ling" (A second order on the salute), etc.[25]

Combined, these statutes provided for an all-encompassing regulation of the cadets' everyday lives, including eating, dress, hygiene, and rituals. The cadets' personal behavior in their daily routines was subject to the control of the academy.

In his speech delivered to Whampoa cadets on May 24, 1924, "Zhuzhong weisheng yu jingshen dikangde daoli" (The significance of sanitation and spiritual resistance), Chiang Kai-shek justified the academy's regulation of the cadets' personal behavior by claiming that in following its rules on personal life, the cadets were protecting not only their own health but also that of their fellow soldiers at Whampoa. Maintaining a healthy body by following the academy's regulations was considered as important as taking the initiative in defending against the enemy.[26] The cadets' ability to perform their daily routines in compliance with these statutes was associated by Chiang with combat readiness. This association justified the academy's policy of controlling the cadets' personal behavior in everyday life. Since the cadet's body belonged to the party state after he entered into the academy, he had to accept regulation of his personal behavior, including eating, drinking, and hygiene.[27] In the same speech, Chiang said, "When you have meals, you should thoroughly clean the dishes and chew the food extremely well. You should also eat food coming from both northern and southern regions. When sleeping, you should buckle the button of the shirt."[28] Chiang even checked the daily sanitary tasks and pointed out the individual companies that had dirty rooms. For Chiang, classroom teaching was not enough for successful training of a model soldier; the soldier should cultivate his military spirit bit by bit in everyday life.[29]

The importance of Whampoa's military discipline in controlling the cadets' personal behavior and daily life was confirmed in the 1982 memoir of Wang Zhuochao (1911–2002), a graduate of Whampoa's tenth class (1933–1936). He recalled that the training at Whampoa was very strict and that there were uniform rules regulating clothing, dining, housing, transportation, and even the proper way of folding the laundry, organizing books, and walking.[30] The academy transformed the cadets into model soldiers by exerting strict regulation over every aspect of their day. The regulation of daily routine strengthened the school's surveillance of the cadets' bodies and cultivated a strong sense of discipline into the minds of the students.[31]

To justify the academy's regulation of the cadets' personal behavior, Chiang made an appeal out of compassion, stressing that the regulation of personal behavior was the manifestation of the academy's care for the cadets and would build positive bonds among them. In the speech "Qingjie jiancha jiangping" (Speeches and comments on the examination of sanitation), delivered on May 16, 1924, Chiang spoke to the cadets: "Our officers pay greater attention to your hygiene than your parents did; and this will create an intimate environment within the academy, which can make you happier even than when you were in your own families."[32] Chiang encouraged the cadets to equate the regulation of body and life by the academy with the care they previously received from their parents. He claimed that the examination of personal hygiene would serve as a vehicle through which the closeness among the cadets could be promoted. The academy's efforts to manage the cadets' daily lives imply that it was attempting to break the cadets' lingering ties to parental authority and establish the academy and Chiang as the ultimate definer of normative cadet behavior. Being highly disciplined by the state was a key characteristic in Chiang's citizenship ideal.

Whampoa's imperative to discipline the soldiers included not only cultivating the cadets' submission to the academy's regulation of daily life but also unifying their understanding and treatment of their bodies. First, Chiang Kai-shek demanded that cadets should have the willpower and moral strength to exert self-control over their bodies. In 1925, when the Academy Army started the Eastern Expedition (*dongzheng*), the campaign against warlord Chen Jiongming (1878–1933) and his supporters, Chiang admonished his cadets that the revolutionary army should have "ten no-fears" (*shi bupa*). "The cadets should not be afraid of death, poverty, coldness, pain, heat, hunger, fatigue, distance, weight, or danger."[33] The academy fashioned the soldier figure in terms of the willpower to resist physical and environmental challenges and risks. Whampoa's civic education, which advocated the self-control of the soldiers' bodies, revealed Chiang Kai-shek's highly disciplined citizen ideal.

Second, the academy celebrated the cult of physical sacrifice among Whampoa cadets. In the speech "Geming junrende renge" (Moral qualities of a revolutionary army soldier), delivered in 1924, Chiang proclaimed: "Our soldiers' only duty and aim is simply to die. If you cravenly cling to life and are scared of death (*tansheng pasi*), not only will you lose the qualification for a soldier, you will even lose your moral quality and therefore cannot be counted as a man."[34] Chiang's speech shows that ever since the establishment of Whampoa in 1924, the courage and willingness to sacrifice the body became a defining virtue not just for a qualified soldier but for every person. The moral quality of a person had become inextricably linked to such traits. Chiang's claim revealed that in training model soldiers at Whampoa he was also envisioning a citizen ideal whose body and mind were disciplined by the state.

To promote the spirit of physical sacrifice among Whampoa cadets, Chiang tried to create a heroic outlook on life and death. Chiang said in a speech delivered in 1924, "In a desperate or hopeless situation, you can survive only if you struggle; and if you die, your death in the service of political doctrine will turn out to be an eternal life."[35] Leng Xin (1900–1987), a cadet in the first class, recalled: "Chiang's teachings always mentioned death, thus forcing our cadets to face it bravely and follow the model of the martyrs. Chiang exalted the meaning of death to a higher level by transforming the gloomy and dark sides of death into a glorious and solemn beauty."[36] In teaching the cadets the cult of physical sacrifice, Chiang advocated a heroic spirit among the soldiers. The soldier figure in Chiang's speeches was depicted as a warrior; combat and death were idealized and romanticized. As Green suggests in his study of Whampoa, Chiang's lecture texts often employed phrases "that were popularly associated with heroic tasks and the knight-errant (*youxia*) spirit."[37] The singing of certain songs promoted by Chiang was also "rich with heroic imagery designed to arouse the cadets' patriotism and their sense of duty."[38] For the cadets who were trained at Whampoa to become model soldiers and potential army officers for the national army, strengthening the body only found its ultimate meaning in its sacrifice. The heroic willingness to sacrifice one's life had become a standard by which both the qualification of a soldier and the moral quality of a citizen were measured. In the discourse on the soldier figure constructed in Whampoa's civic education, the soldier figure became a heroic combat warrior, and the brutality of war and death was overshadowed by this romanticized aura.

Since the heroic spirit to sacrifice one's body was an ideal for a model soldier, even suicide on the battlefield was sanctioned by Chiang Kai-shek as a way for a soldier to assert and fulfill his heroism and duty. In the speech "Qiangde xingzhi yu zuoyong he junren naqiang zhi mudi" (The nature and functions of the rifle and the purpose of holding the rifle for a soldier), delivered to the academy cadets on May 25, 1924, Chiang stated that the weapon's bayonet had a dual purpose: to kill the enemy and as a means of suicide. According to Chiang, "If you are caught by the enemies, it is better to kill yourself than to surrender to the enemies and let them insult the noble bodies of revolutionary soldiers. This is because when you are humiliated by the enemy, so are your nation, your army colleagues, and party comrades."[39] A soldier's body became a symbolic essence of national sovereignty and honor. The physical sacrifice for the purpose of defending the honor of the nation and the party was a glorious death. Since Chiang treated the soldier as the national paragon, his expectations for model soldiers discussed in his speeches and lectures also revealed his vision of citizen ideals.

Suicide under desperate situations in combat was thus glorified as the ultimate way for a soldier to prove his heroism and fulfill his duty. Conversely, suicide in a noncombat situation was condemned as a despicable crime. On September 18, 1925, a cadet named Liang Tianli (?–1925)[40] killed himself after he was punished by the academy by being made to stand at attention (*fazhan*) for ten minutes. On the same day, Chiang delivered the speech "Zisha shi beiqiede fanzui xingwei" (Suicide is a cowardly crime) to comment on Liang's suicide: "Liang's suicide was purely due to poverty and hardship of his family. The economic oppression he suffered destroyed his confidence to survive as a man and finally led to his suicide. He was punished by being made to stand at attention for ten minutes yesterday because of his mistake. Six cadets in total were punished this way yesterday, but why was it that only he committed suicide?"[41] Chiang's commentary on Liang's decision to end his life was that it was the unfortunate manifestation of one man's weakness and vulnerability.[42] He eulogized suicides committed for the sake of national honor, but condemned a suicide committed due to a cadet's inability to handle personal distress. Chiang's different attitudes toward the two kinds of suicide showed his efforts to assert his authority in regulating the soldiers by dominating their understanding of their bodies and lives. Put plainly, to Chiang the soldier was an instrument of war, and as such there were both acceptable and unacceptable ways of disposing of that crucial resource. Both the model soldier and the ideal citizen should, in his view, be politicized and disciplined by being made submissive to the state's domination of their political beliefs, in their daily actions as well as in their understanding and treatment of their bodies.

Chiang's intent of soldier training was not only to control the soldier's body but also to regulate his emotion. Chiang reminded Whampoa cadets in his 1924 speech "Benxiao jiaoyu de fangzhen zhi yan zhi biyao" (Necessity of being strict as an educational principle at this academy) that "your current status is not that of a gentle and fragile (*wenzhi binbin*) scholar (*shusheng*) but that of a gallant warrior (*jiujiu wufu*)."[43] Chiang went on to say that cadets must be able to "bear cruel military training"[44] in order to become such a warrior. This speech proposed two stereotypical male types, the effeminate scholar and the macho soldier, and argued that what differentiated these two groups of men was the toughness to endure strict training. The tenderness of such a gentle and refined scholar was described as a direct contrast to the toughness of a warrior soldier.

Chiang argued that emotional repression was an important quality for a model soldier by condemning crying as a trait of effeminacy. Echoing the view of leaders of other military cultures, crying was deemed a characteristic of incapability of self-control, which deprived a soldier of his manhood. Joshua Goldstein points out that the United States in particular at the time also held strong prejudice against crying within the military ranks. This prejudice was exemplified in observations of related incidents at West Point, where a US Army cadet who cried when severely harassed was labeled unsuitable officer material. It was even inferred that those who cried or could not control their emotions were suspected of possible homosexuality.[45] Chiang justified his regulation of the soldiers' emotions by defining crying as a womanly behavior. In the speech "Keku nailao yu kangkai xisheng zhi biyao" (Necessity of diligence, endurance, and heroic sacrifice), delivered on May 21, 1924, Chiang lectured that "the nature of the revolutionary party is heroic sacrifice; the soldiers should never emulate the mean behavior of crying commonly seen in women."[46] In that speech, Chiang condemned the behavior of crying by citing the example of a cadet named Cheng Ruji,[47] who was put in confinement (*guan jinbi*) as a punishment for his attempts to escape from the academy:

> Today Cheng Ruji was confined in a room. I felt that the air was not good inside the room, and I did not want him to get sick. So I thought that as long as he could repent and start anew (*huiguo zixin*) soon, I would release him. However, this morning he cried loudly in the confinement room. This is extremely detestable. The only moments when a man should shed tears out of emotional pain should be when his nation and his family are destroyed, when his parents die, or when colleagues are killed by the enemy. Any other reason does not deserve crying. You only shed tears when you have absolutely extreme disaster; otherwise crying is the

meanest and the most dishonorable behavior. Our revolutionary soldiers should be real men (*dazhangfu*).[48]

Chiang disciplined the soldiers' emotional expression by limiting it only to their nation, family, parents, and colleagues. By requiring the soldiers not to cry for reasons irrelevant to their nation, family, parents, or colleagues, Chiang advocated the moral virtues of patriotic sentiment, familial loyalty, filial piety, and comradeship as key qualities for a model soldier. In the Nationalist discourse, the soldier figure was not only politicized and disciplined but also morally cultivated. Since Chiang considered the killing of army colleagues by the enemies as one of the few sanctioned reasons for a soldier to cry, the academy advocated fraternity as a definer of the soldier in the emotional dimension.

Forging "Qin, Ai, Jing, Cheng"

Chiang Kai-shek considered fraternity to be an important quality for a model soldier at Whampoa. His advocacy of this was best reflected in the motto of the academy, "Qin, Ai, Jing, Cheng" (Intimacy, Fraternity, Dexterity, Sincerity). This motto was designed to demand both mutual reliance among fellow soldiers and personal loyalty to Chiang. Chiang stressed the bond between himself and his cadets as being the most important type. In the Nationalist discourse on the soldier figure, a model soldier was told to be absolutely loyal to Chiang. Chiang built emotional connections with them to win the soldiers' loyalty.

Chiang started his political career serving as the commandant of Whampoa. He exerted tight control of the academy, trying to use Whampoa graduates to control the army, and using the army to reinforce his political rule. To elicit loyalty from Whampoa graduates, Chiang rewarded them by placing them in important positions. By 1937, for example, thirteen cadets out of the 706 in Whampoa's first class had been appointed as commanders of the infantry corps,[49] which accounted for 15 percent of the commanders in the eighty-seven infantry corps of the Nationalist Revolutionary Army for that year. By 1938, fifty-five cadets from Whampoa's first class had been appointed commanders of infantry divisions, accounting for 24.3 percent of the 226 infantry division commanders.[50] As the historian William Kirby reveals, "When Chiang commanded the Whampoa Academy, its staff members and graduates who went on to achieve government or military posts counted among his most loyal supporters."[51] The faculty and students of the academy, who maintained strong bonds of loyalty to Chiang, were called the Whampoa Clique (*Huangpu xi*), and they held some of Chiang's strongest group ties.[52]

Not only did Chiang place Whampoa graduates in important political and military positions, but he also frequently showed his personal care to individual cadets and built his image as a protector and supporter for his Whampoa students. According to the 1964 memoirs of Zhao Zhen,[53] a graduate of the eighth class (1930–1933) of the academy, after each class graduated, Chiang convened the cadets, calling their names one by one and selecting some of them for individual dialogues.[54] According to the memoir of another Whampoa graduate, Xie Yingbai, written in August 1963,[55] when Chiang found out that a student in the eighth class named Mao Yuli[56] came from his hometown and was also a distant relative of his first wife, he immediately appointed him to join his own bodyguard serving in his official residence (*gongguan*) in Chongqing.[57] Upon the breakout of the Second Sino-Japanese War in 1937, the academy moved to Chengdu in southwest China, five hundred kilometers away from the wartime Nationalist capital of Chongqing. Nonetheless, Chiang still attended commencements after each class graduated.[58] At every commencement ceremony, each graduate was given a sword. On one side of the sword hilt was written the words "From Commandant Chiang," and the bottom of the blade was carved with the phrase "We win, or we perish" (*chenggong chengren*). The sword became a symbol of Whampoa graduates.[59] Chiang's treating the cadet who came from his hometown favorably and his giving a special sword carved with his name to each Whampoa graduate showed that Chiang showed favoritism to those who had personal connections to him. Since Chiang trained Whampoa cadets to become model soldiers and army officers for the national army, he hoped that the Whampoa cadets could help reinforce the loyalties of the national armies to him.

According to the memoirs of Zhao Zhen, Whampoa maintained a communal dining (*gongcan*) system in which student representatives and faculty above the rank of major met for dinner every Saturday. During the dinner, after the academy dean gave a speech, the students took turns telling jokes and stories and singing operas.[60] According to the Whampoa graduate Qi Xiangming's memoir, written in June 1963,[61] on holidays the academy often invited Chiang and his wife to have dinner with Whampoa cadets.[62] This communal dining system was designed not only to create the impression of internal unity and close bonding horizontally among Whampoa cadets but also to strengthen the cadets' vertical loyalty to Chiang by strengthening their emotional attachment via his interaction with their everyday lives.

The academy not only cultivated the personal loyalty of the cadets to Chiang, but it also advocated close bonding among the cadets themselves. As Andrew Scobell comments in his study of Chinese military forces, "[In Whampoa] while the rhetoric was of selfless dedication to the Chinese people, in practice the stress was on proximate loyalties to military superiors and fellow cadets."[63] The cadets who were trained at Whampoa as model soldiers and potential army officers needed to possess a strong identity with the academy and a close connection with their fellow cadets.

The bonding among the academy cadets was institutionalized by the military edict *Lianzuofa* (Law of joint responsibility) that was implemented at Whampoa in 1925. This edict, which was later introduced to the Nationalist army, demanded punishment by death for those who retreated from battle without permission. It aimed to ensure the maintenance of military discipline by binding the destiny of soldiers together and making them willing to live and die together. The details of the law were as follows:

> If the squad leader (*banzhang*) retreats with the whole squad, the squad leader shall be killed; if the platoon leader (*paizhang*) retreats with the whole platoon, the platoon leader shall be killed; if the company commander (*lianzhang*) retreats with the whole company, the company commander shall be killed; so it is with the battalion commander (*yingzhang*), the regiment commander (*tuanzhang*), the division commander (*shizhang*), and the army commander (*junzhang*). If the army commander does not retreat but the whole army retreats and the army commander dies as a consequence, the division commanders in this army shall be killed; if the division commander does not retreat but the whole division retreats and the division commander dies as a consequence, the regiment commanders in this division shall be killed. The same applies to the battalion, the company, and the platoon. If the squad leader does not retreat but the whole squad retreats and the squad leader dies as a consequence, all the soldiers in the squad shall be killed. After the implementation of this *Lianzuofa*, everyone in the

army should feel a knife hanging above his head (*dao jiazai bozi shang*) or a rope tied around his feet (*sheng jizai jiaoshang*).[64]

Implementing *Lianzuofa* revealed the GMD's state-building agenda. The bonding among the soldiers that was legalized by this military edict meant both mutual dependence and mutual surveillance for survival. Thus, this edict was a tool that the GMD employed to complement its efforts to discipline the soldiers.

The *Lianzuofa* edict also reinforced the hierarchical relationship among the soldiers. According to the 1975 memoir of the Whampoa graduate Du Congrong (1902–1979), Chiang told his cadets: "The soldiers had to listen to the order of the higher commander for the action. Whoever retreats from the battlefield against the order shall be killed in compliance with *Lianzuofa* even though he survives from the combat against the enemy."[65] As Chiang said in his speech "Benxiao zhi shiming yu geming de rensheng" (The mission of this academy and the revolutionary life), delivered on May 8, 1924, "The lives of all the cadets in Whampoa are ultimately one life."[66] The bonding, which was based on mutual dependence and surveillance as well as hierarchical relationships, served as an institutionalized part of the Whampoa military culture through its military edict and the academy's motto.

To strengthen bonding among the cadets, Whampoa set up a series of rituals that required the participation of all the students. The most important and most frequently observed ritual was the Premier Sun Yat-sen weekly memorial service (*Zongli jinian zhou*). The Nationalist Party Central Committee in Guangzhou first decreed this remembrance in 1926 for all party and government officials.[67] The weekly assembly was fully instituted by the Nationalist regime in all government offices as well as in all the schools, colleges, and universities under its effective jurisdiction by 1927.[68] Every Monday morning the academy came together to sing the party anthem, bow three times toward Sun's portrait, recite his Last Testament (*Zongli yixun*), and observe three minutes of silence before the portrait.[69] The assembly was intended to "provide an occasion for the students to receive moral lessons, guidance on academy activities, analyses of political events, reports on current affairs by academy administrators, public figures, and party leaders, as well as lectures on academic subjects by prominent intellectuals."[70] By requiring Whampoa and other government offices and social institutions to observe this ritual, Chiang intended not only to produce army officers but also to promote the disciplined citizenship ideal within the army.[71]

In order to reinforce the cadets' bonds with the academy, Whampoa not only required them to attend the weekly memorial service but also obliged them to observe holidays to commemorate watershed events that shaped the GMD's history. The holidays that were observed at Whampoa included the Commemoration of Sun Yat-sen's Death (*Zongli shishi jinian ri*, March 12), the Commemoration of the Huanghuagang Martyrs (*Huanghuagang lieshi jinianri*, March 29), Labor Day (*Laodongjie*, May 1), the Commemoration of National Shame (*Guochi jinianri*, May 9), the Commemoration of the Shanghai Massacre (*Shanghai can'an jinianri*, May 30), the Commemoration of the Foundation of the Academy (*benxiao kaishi jinianri*, June 16), the Commemoration of the Shaji Massacre (*Shaji can'an jinianri*, June 23), the Commemoration of the Capture of Party Representative Liao Zhongkai (*Liao [Zhongkai] dangdaibiao beibu jinianri*, August 20), and the Commemoration of Premier Sun's Birthday (*Zongli danchen jinianri*, November 12).[72] These holidays were all centered on the themes of Sun Yat-sen and the Nationalist revolution. The academy conceived the remembrances and the observation of holidays as ceremonies to unite the minds of the cadets and build a shared identity among them with the academy, the Nationalist army, and the Republic of China government.

Rituals of observing political holidays and attending the weekly memorial services show the lasting legacy of Confucian culture at Whampoa Academy. The classical Confucian text *Liji* (Book of rites) placed ritual (*li*) rites—ceremonies, manners, and deportment that distinguished Chinese from other ethnic groups—at the heart of the state. As Patricia Ebrey comments in her study of family rituals in imperial China, "Ritual (*li*) became a central concept in Confucian thinking about human nature, ethics, social harmony, cultural identity, and the relationships between the human world and the sphere labeled heaven."[73] The Confucian tradition asserted that social order and human virtues could be attained through observing rituals. The stress on rituals at Whampoa reveals that the training of a model politicized soldier in modern China was shaped by Confucian ideals and concepts.

The commemoration of rituals and holidays was not the only important technique that Whampoa utilized in promoting unity among the academy cadets. Another technique the academy adopted for this purpose was the establishment of an alumni association—an engagement system that first appeared in Western universities in the nineteenth century. The Whampoa Alumni Association (*Huangpu tongxuehui*) was founded on June 27, 1926, and was organized by the academy's political department. Chiang served as the president of the association. In a speech to the academy cadets delivered in November 1926, Chiang claimed that "to oppose the alumni association is to oppose Chiang himself."[74] The ultimate goal of the association was to follow the Last Testament of Premier Sun Yat-sen and pursue the nationalist revolution. The mission of the association was "using the Whampoa Academy as a network center to build emotional attachment among the graduates, to encourage and support each other, to unify the spirits and minds of them."[75] According to the rules set by the association in 1928:

> Cadets and graduates from every class of the academy all automatically become members of the association. The association is in charge of recording and checking the members' performance, appointment, removal, promotion, and transferal. No matter whether the cadets have graduated or not, they are all subject to surveillance and direction by the association and shall pay loyalty to the Nationalist Party and the commandant of the academy, practice the Three Principles of the People, and shall not join any other political organization. Any cadet or graduate who breaks the rules is subject to severe punishment or will be considered as a rebel against the nation.[76]

In 1930 the association changed its name to the Investigation Office of Whampoa Academy Graduates (*Huangpu junxiao biyesheng diaochake*). In theory, it was the headquarters where Whampoa graduates could build connections and bond. However, in practice, it took charge of all aspects related to the careers of Whampoa alumni, including registration, investigation, promotion, appointment, and punishment. The association thus served as an agent for Whampoa to consolidate its discipline of the cadets even after they graduated.[77]

The memoirs of Whampoa graduates show that strong unity among the cadets did develop in the academy. According to the 1963 memoir of one academy graduate, Qi Xiangming, "[At Whampoa] only the academy graduates played the managing roles (*zhuren*); faculty and staff who had not studied at Whampoa were nothing but employees in the eyes of the cadets."[78] Based on Scobell's research, "the 7,399 cadets in the five classes that graduated from the Whampoa Academy between 1924 and 1927 established strong bonds of personal loyalty to their classmates. Whampoa cadets were instilled with a strong sense of *esprit de corps* that has been called the 'Whampoa spirit' (*Huangpu jingshen*)."[79] The establishment of the alumni association in 1926 to promote unity among Whampoa graduates revealed Chiang's intention of disciplining the soldiers and reinforcing his authority in the army for the long term.

Confronting the GMD's Heroic Soldier Ideal

The analysis above has shown that civic education at Whampoa Academy constructed a discourse on the soldier figure that included a heavily defined code of military ethics. The key qualities that this code denoted were strong political indoctrination, absolute loyalty to Chiang Kai-shek, submission to military discipline, and hierarchical bonding among the cadets. The academy trained the cadets to become model soldiers by cultivating and reinforcing this code among them. Training model soldiers at Whampoa revealed Chiang's intention of creating politicized, disciplined, and morally cultivated citizens in the nation. Since Whampoa was founded to train potential army officers for the national army and raise the professional standards of the larger army, Chiang expected Whampoa cadets to reform national armies by extending this code to them. However, the memoirs of Whampoa graduates suggest that in reality the soldier ideal deliberately constructed by Chiang often met resistance from Whampoa cadets and regional warlords.

Although the academy advocated strict adherence to military discipline as a key quality of a model soldier, some Whampoa cadets believed that the Nationalist ideal of defending one's hometown was more important than complying with Chiang Kai-shek's nonresistance policy. After the Mukden Incident on September 18, 1931, when Japan invaded northeastern China, Chiang emphasized that, at a time of domestic turmoil and inadequate preparation, China must avoid an all-out war with Japan. However, Chiang's *annei rangwai* policy (first settle the country internally, then resist the invader) was unpopular. As Hans van de Ven comments, "For any government to be shown to be weak cannot but damage its reputation and the refusal to stand up to an aggressor provides easy openings for its critics."[80]

Chiang Kai-shek's *annei rangwai* policy was even questioned by the cadets he commanded at Whampoa, who were supposed to be unconditionally loyal to him and his ideas. The ninth class of the academy enlisted over one thousand students, two-thirds of whom came from northeastern provinces that were quickly occupied by the Japanese.[81] Many Whampoa cadets asked to suspend their studies and go back to their hometowns to fight against the Japanese. These cadets shouted slogans such as, "It is soldiers' humiliation not to recapture the lost territory" and "We cannot forget to save the nation when pursuing our military studies; to save our nation is the ultimate purpose to pursue our studies."[82]

According to the memoir of Whampoa graduate Xie Yingbai (1890–1969), written in August 1963, a cadet from Liaoning Province named Li Yihu[83] insisted on returning to his hometown to resist the Japanese invasion whether the academy approved or not. He even claimed that he would kill himself with a knife if the academy did not approve his request. Zhang Zhizhong (1890–1969),[84] who served as dean of the academy from 1929 to 1937, did not punish him. Instead, Zhang praised him for his patriotic sentiments and allowed him and ten other students to return to Liaoning.[85] Li Yihu's request to fight against the Japanese was not in accordance with Chiang's policy. His threat to kill himself if his request was rejected was also in conflict with Chiang's condemnation of suicide in noncombat situations. However, patriotic fervor encouraged him to voice discontent with Chiang's policy and at the same time saved him from being punished by the academy.

The opposition of some Whampoa cadets to Chiang's policy of temporary appeasement is also revealed in Lincoln Li's study on student nationalism. Li notes that some Nationalist military officers attempted to rekindle an ethic of patriotism and national service within the increasingly bureaucratized and demoralized Nationalist party-state in the early 1930s. They were able to express the fiery anti-Japanese sentiments at a time when Nanjing was still pursuing a policy of appeasing Japan.[86]

Other memoirs reveal that in the eyes of many Whampoa cadets, following military discipline should not be absolute. Sometimes they resorted to violence in their demonstrations against the unfair aspects of the academy's administration. After each class graduated, the academy would distribute funds for travel and clothes to each graduate. Whampoa authorities and fiscal departments often took the opportunity to skim off a profit for themselves. According to Whampoa graduate Cheng Tingrong's memoir published in May 1963,[87] the cadets' dissatisfaction with the administrative corruption in the academy eventually led to a riot among the graduates of the seventeenth class. These graduates believed that the standards that Whampoa used to calculate traveling expenses were unfair. They requested that their traveling reimbursements be based on the actual routes and periods of travel.

After their request was denied by Xiao Zanyu (1905–1999), the director of the political department, the students were so upset that they broke into Whampoa's accounting offices. They even broke into the houses of the directors of these offices because they were annoyed by what they regarded as the luxury inside the houses. After this riot, the academy caught about one hundred students. However, because of Chiang's protective attitude toward his cadets, only seventeen of them were sent to prison, and sixteen of those were released within one year. Cheng Tingrong's memoirs do not make clear to what extent the cadets who participated in this riot were able to force the academy authorities to adjust their policies for distributing funds. But the riot did show that the cadets would not shy away from violence in their protests against the academy's authorities for something they viewed as unfair.[88] Chiang's efforts to discipline soldiers were challenged even by his own cadets.

Although Chiang attempted to forge cadets' emotional attachments to him at Whampoa, loyalty to him was not the only emotional resource that guided socialization among the students. Schoolmateship (*tongxue qing*), comradeship (*zhanyou qing*), and native-place sentiment (*tongxiang qing*) all served as bases for bonding networks among Whampoa cadets. The cadets took advantage of connections in various forms to broaden their networks and strengthen their groupings. Loyalty to Chiang did not entirely occupy the emotions of the cadets; lateral bonding based on various private relations also contributed to their socialization in the academy environment.

The emotional bonding among the cadets through diverse networks sometimes encouraged them to confront the academy authority to protect fellows in their groups. The Whampoa graduate Tan Dingyuan (?–2008) provides an example in his memoir, written in November 1961. In the winter of 1935, says Tan, Chiang held a large-scale military drill near Nanjing. All the cadets in Nanjing participated in the drill and were assigned to different army units. The engineering company of the 88th Division failed to prepare all the required equipment and to build a military bridge in time;

their mistakes led to the interruption of the drill. The commander of the sapper company was criticized and then committed suicide. At a meeting following the events, the dean, Zhang Zhizhong, spoke highly of the commander's suicide and reemphasized the importance of military discipline. However, cadets from the commander's hometown showed dissatisfaction with Zhang's attitude and expressed their sympathy toward the commander. They even brought the commander's wife and two children to the meeting, letting them express the grief that was caused by the death of the commander. Eventually, Zhang announced at the meeting that the academy would provide some funds to the commander's dependents. The cadets who participated in the drill also made a donation to the commander's family.[89]

The cadets did not unconditionally embrace the soldier ideal constructed by Chiang. Instead, they sometimes protested against what they viewed as unacceptable administration at Whampoa. These events reveal that tensions existed between Chiang and the cadets. Moreover, the designation of Whampoa cadets as model soldiers in the GMD's political discourse was often resisted by society, particularly provincial warlords. Chiang's efforts to extend the politicized and disciplined soldier ideal to the national army in order to restore the tarnished image of the military and soldiers and raise the professional standards of the larger army were compromised.

Republican Chinese society did not wholly accept Whampoa cadets as model soldiers; nor did they retain belief in the army. Whampoa cadets did not always receive a positive reception from civilian society. The newspapers often reported events involving cadets fighting at theaters, train stations, and restaurants. Ordinary people in Nanjing even referred to Whampoa cadets as "locusts" (huangchong).[90] In 1934, Chiang selected four hundred students from the academy's tenth class and sent them to mingle in a crowd of civilian students who were participating in a protest. Chiang's intention was to let the cadets produce some chaos among the students in order to break up the student protest. Afterward, Chiang did not respond to questions and criticism from the public concerning the cadets' having been sent in to disrupt the protest.

The signal that the cadets received from this event, according to Tan Dingyuan's memoir, was that as long as what the academy cadets did was approved by Chiang, they would get favorable treatment even though their actions were criticized by the public.[91] In brief, Chiang took advantage of the cadets' loyalty to him by using them to forward his personal short-term political goals. In doing so, Chiang failed to repair the tension between the soldiers and the civilians and regain the society's faith in the military. The privilege Chiang gave to his loyal soldiers was also criticized by the Nationalist army officers. The historian Frederick Fu Liu, who served as a Nationalist army officer during the Second Sino-Japanese War, criticizes the undue influence of Whampoa Military Academy graduates in the political life of the Nationalist regime and the tendency to let personal associations take priority over political considerations. Liu attributes much of the difficulty encountered by the government of Chiang to these factors and to the obstacles created for effective civilian rule.[92]

Not only did the cadets have conflicts with the civilian community but they also were considered major competitors by common soldiers in regional army units. Historian Eugene Levich's study on Guangxi warlords in the 1930s reveals the hostility from the Guangxi forces toward the cadets in Chiang's Whampoa Academy. The major warlord in Guangxi, Li Zongren (1890–1969), claimed that Chiang's partiality toward graduates of the Whampoa Military Academy, and toward the military units under his direct command, undermined morale in the other GMD forces and undermined discipline in Chiang's own forces.[93] Li recalled that during the Northern Expedition he often heard complaints about Chiang's partiality toward his own First Army from high-ranking officers of the Second, Third, Fourth, Fifth, and Sixth Armies.

Levich cites several examples from Li Zongren's 1958 memoir on his experience during the Northern Expedition to show Li's dissatisfaction with the extension of special privileges to Whampoa graduates. For example, when sandals were issued to the armies, by Chiang's order each man of the First Army got two pairs; this left some men in the other armies receiving none. Whenever a middle- or lower-ranking officer of the First Army who came from the Whampoa Academy squandered pay by gambling or the like and confessed his mistake to Chiang, Chiang not only failed to punish him, but he also allowed the man to draw needed money from the Military Supply Department.[94]

The hostile attitudes of common soldiers to Whampoa cadets also sometimes led to violence. One such example is provided in Tan Dingyuan's memoir. After the Shanghai Incident on August 13, 1937, there was a significant loss in the number of lower-level army officers. The academy dean, Zhang Zhizhong, asked the academy to select brave cadets from the eleventh and twelfth classes to join the battle. The death ratio for these cadets at the front was very high, and many of them were found to have been shot in the back. But they weren't shot by the enemy while retreating. The explanation for this phenomenon, according to a Whampoa graduate's memoir, was that ordinary army soldiers expected to get promoted after their commanders were killed by the enemy during battle. The seconding of Whampoa cadets into their units disturbed their promotion plans, causing resentment. The Whampoa cadets were shot by the soldiers they led.[95] This example shows that, in reality, Whampoa cadets who were assigned to local army units were not always highly respected and seen as model soldiers like Chiang had hoped. Instead, they were sometimes viewed by ordinary army soldiers as competitors and treated with hostility or even violence. The politicized and disciplined citizen-soldier ideal advocated in Chiang's speeches at Whampoa was not embraced by all in the national army.

Another factor that caused tension between Whampoa cadets and ordinary soldiers was that Chiang did not have uniform or sole direction of the national armies, and thus central authority was rivaled by provincial army forces. As the historian Edward McCord comments, despite the victory of the Nationalist Party's Northern Expedition and the reunification of China under a new Nationalist government in Nanjing in 1927, warlords were not simply eliminated. "Many were absorbed into the Nationalist Party army and, in exchange for their allegiance to the new government, allowed to retain a considerable degree of political autonomy in their garrison areas."[96] The state-building strategy employed by Chiang Kai-shek to expand his power deeply into the regional army forces involved opening temporary training classes or branch campuses affiliated with Whampoa Academy. In the eyes of the provincial warlords, however, Whampoa was not a national military reform vanguard to train model soldiers and potential officers; it was Chiang's personal political tool to control non-central army forces. The heroic aura surrounding the soldier figure constructed by Chiang was therefore resisted by the warlords.

After the January 28 Incident in 1932 (also called the First Shanghai Incident), the academy at Nanjing became the target of bombing by the Japanese, which made it impossible to continue the regular operation of the school. Chiang Kai-shek decided to build a branch campus at Chengdu in Sichuan in 1935 and sent the academy dean, Li Minghao (1897–1980), to serve as the director of the Chengdu branch campus. Saving the academy from the Japanese bombing was only one reason for the move; Chiang also intended to expand the power of his central army into the armed forces of the regional warlords. The Chengdu branch campus offered officer training classes to surplus officers (*bianyu junguan*) from the armies of Sichuan warlords Liu Xiang (1888–1938) and Liu Wenhui (1895–1976).

After a six-month training period, these cadets were assigned to units that comprised the Sichuan armies to serve as lower-level officers (usually below the rank of battalion commander). The campus continued enrolling lower-level officers from the Sichuan army who had not received military education. Officially, the classes were intended to unify national military education and to help the Sichuan armies train their middle- and lower-level army officers in the use of new weapons; both excuses made it hard for the Sichuan warlords to refuse.

However, the true purpose of these classes may have been to break up the Sichuan warlord armies by building a foundation for pro-Chiang sentiment among the military officers in those armies. Therefore, the Sichuan armies and this Chengdu officer training program often had conflicts. According to the memoir of Qian Daquan (1908–?), compiled in 1963,[97] when the cadets in the Chengdu army officer training class heard that Liu Xiang intended to dismiss the class, they protested against this decision by threatening to attack Liu's official residence.[98]

The establishment of the Luoyang branch campus in the summer of 1933 is another example that shows Chiang Kai-shek's intention not only to train model soldiers but also to cultivate his trusted fellows among non-central armies. The Luoyang branch campus recruited platoon and company leaders from non-central armies into a ten-month military officer training class. They were sent back to their army units after the training. However, according to the October 1963 memoir of Whampoa graduate Zheng Dianqi (1907–?), some regional warlords, such as Yan Xishan (1883–1960) in Shanxi, Han Fuqu (1890–1938) in Shandong, and Liu Xiang (1888–1938) in Sichuan, simply refused to accept the cadets who were sent to their armies. Other warlords adopted discriminatory attitudes toward their old subordinates who came back after their training at the branch campus of Whampoa, and tried to elbow them out. Many cadets were not given important positions after they returned to their original army units and, therefore, were not able to bring the pro-Chiang sentiments into the regional armies as Chiang had expected.[99] Chiang's agenda to discipline the soldiers by constructing a close hierarchical bonding among the army, with him as the leader, was hindered by these rivalries. The hostile and prejudicial attitudes of regional armies toward Whampoa graduates hindered Chiang's state-building efforts to use the Whampoa Academy as a vanguard to raise the professional standard of national armies.

Chang Rui-te provides additional reasons for why the role of the Whampoa Academy in promoting the combat capacity of the national army should not be overestimated. For example, he explains that the academy could not expand the school quickly enough to meet the demand for junior officers in the ongoing anti-Communist campaigns.[100] With the outbreak of the Second Sino-Japanese War in 1937, "the high number of casualties and the rapid expansion of the army essentially prevented Academy graduates from exerting any decisive influence on the quality of Chiang's troops."[101] In addition, the demand for new officers grew so quickly in the first years of the war that the threshold entry requirements were lowered to bring in more candidates.[102] Chiang's state-building agenda of training model soldiers was resisted by Whampoa cadets themselves and also undermined by provincial warlords. In the end, the brutal reality of the Second Sino-Japanese War and the worsening situation at the front in the first years of the conflict impeded Chiang's state-building agenda of military reform.

This chapter has analyzed the Nationalist discourse on the soldier figure that was constructed using civic education at the Whampoa Military Academy in the early years after Whampoa's establishment in 1924. The examination of Chiang Kai-shek's speeches delivered at Whampoa in 1924 shows that ever since Whampoa's establishment, Chiang tried to teach a code of ethics to Whampoa cadets and train them to be model soldiers and potential army officers for the national army. The code was designed to cultivate the soldiers' obedience to the political doctrine of the Three Principles of the People, their submission to the discipline and regulation of personal behavior and emotional expression, and their subordination to the hierarchical bonding in the army, with Chiang as the indisputable leader. This code allowed the Nationalist state to convert a soldier into the image of a heroic warrior who considered death in combat a glorious honor and was tough enough to restrain his emotional expression.

The building of Whampoa and its civic education system revealed the GMD's state-building agenda to use Whampoa as a political tool to reinforce the legitimacy of the regime. Chiang tried to train politicized, disciplined, and morally cultivated model soldiers at Whampoa; in doing so, he also envisioned a politicized, disciplined, and morally cultivated citizenship ideal. After Whampoa cadets graduated and became officers in the national army, Chiang expected them not only to raise the professional standards of the national army but also to extend such citizen-soldier ideals to the larger army.

The memoirs of Whampoa graduates written in the 1960s reveal that Chiang's intentions were not fully achieved since the image of the soldier figure constructed by Chiang was confronted or resisted by some Whampoa cadets and provincial warlords. The cadets employed a nationalistic appeal to confront Chiang's policy of temporarily appeasing Japan. Loyalty to Chiang was not the only element that shaped the cadets' identities and emotions. In addition, they developed private social networks that were based on various emotional connections. Chiang advocated fraternity among the cadets to strengthen his authority and the influence of his Whampoa clique in the army and government, but that same fraternity enabled the cadets to protest against what they viewed as unacceptable practices by the academy's administration. Nominally, Whampoa cadets were trained to become model soldiers for the national army, but they were not always respected or favored by ordinary soldiers and provincial warlords; in fact, they often received prejudiced, hostile, or even violent treatment from them.

The building of Whampoa after 1924 was only Chiang's initial state-building effort to expand the state institution and create a politicized, disciplined, and morally cultivated citizenry. The different treatments of the political discourse of the soldier figure constructed by the GMD and particularly by Chiang reveal the existence of tensions among the GMD state and Whampoa cadets, provincial warlords, and civilian society. In the 1930s, as the threats from the Chinese Communists and the Japanese increased, Chiang consolidated his state-building project by expanding state institutions and extending the state's influence into local societies. Chiang's efforts resulted in the issuing of the first compulsory conscription law in the mid-1930s. In promulgating this law, Chiang tried to extend the disciplined, politicized, and morally cultivated citizenship ideal that he originally envisioned in the training of his model soldiers from Whampoa to the rest of society. We will turn to this topic in the next chapter.

2

Enlisting Citizens in the Military Mobilization of the Nationalist State

The Nationalists' goals of creating a disciplined and politicized army and building a strong, unified nation required the GMD government to maintain military mobilization and ensure sustained recruitment for its army. To this end, in 1933 the GMD issued what became the first compulsory conscription law of modern China. This chapter approaches the project of constructing the soldier figure using military laws, soldier education materials, and political propaganda from the 1930s and 1940s. The compulsory conscription laws, army textbooks, and New Life Movement propaganda to be discussed in this chapter represent aspects of the GMD's state-building efforts. Together, they reveal the Nationalist Party's intention of extending the militarized, politicized, disciplined, and morally cultivated citizenship ideal that Chiang Kai-shek envisioned at Whampoa.

The soldier was glorified in Nationalist military law and in political propaganda as a national hero and paragon. He was postulated as a model citizen and the epitome of morality, a person who should be emulated by society. The goals of the GMD state behind such rhetoric were twofold. First, it aimed to achieve military mobilization by fostering the military spirit and encouraging society to follow the conscription law and perform obligatory military service. Second, it worked to strengthen its state-building project by creating a national identity among the people as militarized, politicized, disciplined, and morally cultivated citizens. However, the real-life conditions of the recruited conscripts did not effectively attract society to the army. The public still viewed the soldier's life as miserable. Negative attitudes prevailed toward the conscription system. The GMD's state-building and military mobilization goal of elevating the soldier's status through its rhetoric met resistance from society, just as it had from warlords and even Whampoa cadets.

The Soldier as a Model Citizen

By the time the Nanjing Nationalist state was founded in 1927, the majority of the GMD's military forces were composed of long-serving regulars recruited for pay (*mubing*).[1] They had been born to poor families and came from the warlord armies. In her discussion of Chinese common soldiers from the period of 1911 to 1937, Diana Lary remarks that soldiers enlisted with whatever unit recruited them or would take them. "Most were straightforward mercenaries, some found themselves fighting for a cause, a principle added to their wages."[2] According to a 1932 survey of a Taiyuan army unit of 946 soldiers carried out by the sociologist Tao Menghe (1887–1960), 73 percent of the soldiers claimed that their family finances were poor, 24 percent of them claimed that their income barely made ends meet, and only 3 percent of them claimed that their families managed to save a small portion of their earnings.[3]

Armies in the Warlord Period (1916–1928) prior to the establishment of Chiang Kai-shek's Nationalist state were not controlled by a unified bureaucracy. As Hans van de Ven notes, "Systems of recruitment were neither national nor bureaucratized. Nor were the armies themselves well integrated."[4] However, this situation began to change during the period that encompassed the Nanjing decade of 1927–1937 and through the eight years of the Second Sino-Japanese War. The GMD government issued the compulsory conscription law in 1933, put it into implementation in 1936, and made several revisions during the war.

The advocacy of compulsory conscription in modern China was rooted in the historical background of the last years of the Qing dynasty (1644–1911), when the military took on new importance in China. As Colin Green suggests, many politically engaged Chinese intellectuals, such as Gu Hongming (1857–1928) and Liang Qichao (1873–1929)—who were driven by fear that China was on the verge of becoming a "lost country" (*wangguo*) like Poland or Vietnam—took up the cause of militarism, arguing that to survive in a hostile world the Chinese people would have to overcome their spiritual malaise and lack of martial vigor.[5]

The efforts of political and military authorities to implement the compulsory conscription system started as early as the birth of the Republican era in 1911. As Lary shows, Yuan Shikai (1859–1916), who was elected provisional president of the Republic of China on February 14, 1912, floated a compulsory conscription scheme and issued formal plans to that effect in 1915. That particular effort did not sustain itself, as it was scrapped with Yuan's death a year later.[6] Xu Chonghao (1882–1959), a member of the Chinese Revolutionary Alliance (*Tongmenghui*) who had participated in the 1911 Xinhai Revolution, published in 1929 the *Zhengbing zhi yange ji shixingfa* (The transformation of the military recruiting system and its implementation), advocating the implementation of the conscription system. Xu believed that every citizen should possess military knowledge in order to strengthen the nation's ability to resist foreign threats. Comparing the conscription system with the mercenary system employed by warlord armies, Xu argued in strong favor of the conscription system by remarking on its advantages:

> Given the spacious size of China, the conscription system would allow China to enlarge the military forces while saving expenditures. Under the conscription system, all able-bodied men (*zhuangding*) take turns in serving the army and retire from the army after a certain period. With a set number of conscripts each year, the military expenditure could be prepared in advance in the state's budget. However, under the mercenary system, soldiers are recruited only for pay and might serve in the army for all his life. If the state needs to enlarge the army, it has to increase the expenditure for the new recruits. In the emergence of a national crisis, the state is usually in need of a number of soldiers who have military knowledge, but the mercenary system might make the state unable to afford recruiting enough soldiers.
>
> The conscription system would also allow the army to better maintain military discipline. The soldiers recruited for pay are mostly jobless vagrants (*wuye youmin*), or even robbers and bandits (*daozei tufei*) with complicated backgrounds. Men with some knowledge and coming from decent families all consider military service to be shameful. Those soldiers recruited only for pay are reluctant to follow military discipline. They often resort to aggression and violence and do all kinds of evil things by taking advantage of their weapons. However, under the conscription system, the conscripts consist of people who have basic knowledge and come from decent families. They have self-respect and do not dare resist the law and military discipline.[7]

Xu's understanding of the conscription system was more a criticism of the mercenary system employed by the warlord armies. He did not associate military service with national duty as a citizen; nor did he provide a thorough legal framework on the implementation of the system. However, as a military general, Xu did propose recruiting all able-bodied men as the one way to reform the army and build a strong nation.

Van de Ven reveals in his study on war and nationalism that He Yingqin (1890–1987), who served as chief of staff for military affairs in the GMD government from 1930 to 1944, proposed a system of national military service in his 1928 draft "Proposal to Change National Military Service." The reasons He gave in favor of a national military service system were similar to Xu Chonghao's claims. He particularly pointed out that national military service was a characteristic of both economically advanced countries in the West like Germany as well as of revolutionary countries like France. Following the Japanese model, He outlined the structure of the national military service as the combination of a standing army and ready reserves.[8]

The Nationalist state reintroduced the traditional local control system of mutual surveillance (*baojia*) between 1932 and 1934, and it served as the basis for He's outline for the national military service system. This system entailed the registration of all households, and its structure was as follows: "Groups of ten or so households were formed in a unit termed a *jia*. Ten or so *jia* became a unit known as *bao*. Several *bao* were grouped together in 'associated bao' (*lianbao*) at the district level. The *baojia* required households after registration to sign mutual responsibility."[9] Theoretically, *jiazhang*, who headed one *jia*, and *baozhang*, who headed one *bao*, each were elected by popular vote from *jia* and *bao* residents, but "many *bao* and *jia* leaders were *de facto* appointed to their positions by higher authorities."[10] As the historian Jae Ho Chung comments, the leaders of the *baojia* organizations "acted more as government functionaries than civic-minded elected officials responsive to popular demands."[11] The *baojia* organizations also had military functions, which were to register and administer military training of able-bodied men who were potential conscripts, and to organize local militia groups to cooperate with the government in preventing popular insurgent activities.[12] The *baojia* system did not promote self-government at the local level. Indeed, it was a result of the GMD's project of expanding the state institutions and projecting the state's influence into local society. Chiang Kai-shek hoped that this system could militarize the grassroots organizations.[13] As noted by Van de Ven, the *baojia* was intended by the GMD authorities to serve as the basis for a compulsory military service system.[14]

The introduction of *baojia* organizations enabled the Nationalist state to introduce European-style conscription by the early 1930s.[15] Its efforts resulted in the formulation of the first compulsory conscription law (*Yiwu zhengbingfa*), which was issued on June 17, 1933, and was to be implemented on March 1, 1936. Based on this law, the compulsory conscription system was initiated not to replace but rather to complement the preexisting mercenary system. The first item of the law proclaimed that "to perform military service is the national duty of the males in the Republic of China."[16] All males between eighteen and forty-five were subject to the draft. All males after their eighteenth birthday were obliged to enroll in citizen-soldier (*guominbing*) units. Citizen-soldiers were trained at the district level on basic military skills, including weapon use, fortification construction, execution of basic orders, reconnaissance, and liaison. The training for citizen-soldiers also included civic education "delivered in the form of lectures on the Three Principles of the People, the New Life Movement, and National Economic Construction."[17] Able-bodied men between the ages of twenty and twenty-five were to be recruited into the army to perform active duty (*xianyi*) for three years. Soldiers who retired from active duty were then expected to serve the military as reservists (*zaixiang junren*) until they reached the age of forty-five. During the agricultural off-season, citizen-soldiers and reservists were expected to attend military training, drills, and reviews at local levels on a regular basis, perform policing services during temporary emergencies, and be subject to reactivation during times of war.[18] Military service required for citizens was explicitly gendered. According to the law, it was only men's national duty as citizens to join the army and defend the country; women did not have any duty of military service.

The 1933 compulsory conscription law was one of the GMD's state-building efforts to reform the military and to discipline and militarize society. As Van de Ven comments, the purpose of the military service law was to raise awareness of the nation as well as to nurture heroic and martial attitudes.[19] By demanding military service as the legal duty for national citizens, the GMD was creating a militarized ideal of citizenship. Legalizing military service as a national duty for citizens was intended to build the concept of citizenship among the public during the period of political tutelage (*xunzheng*, 1928–1947).[20] The vision that the Nationalist state attempted to forge among the public was that military service was a prerequisite for a man to become a full-fledged citizen.

The link between military service and citizenship status was reinforced in the Nationalist legal document *Zhonghua minguo xianfa cao'an shiyi* (Illustration on the Republic of China Constitution draft), published in 1936. It stated: "Since the nation cannot survive in the world without military forces, to perform military duty is the natural obligation for the people. Although every nation state employs different military systems, the general principle is the same—which is that citizens (*guomin*) have the duty to serve the military. Modern warfare is a national cause that requires military training of national citizens in peacetime to handle emergency situations."[21] According to this legal document, since the military constituted the essence of a nation's security and power, serving in the military thus became a national obligation for every male citizen. Therefore, performing military service was considered by the GMD as the vehicle by which all men could develop a unified national identity regardless of class, age, profession, family, or education. They would be shaped into the ideal image of military-ready and disciplined citizens.

The Nationalist state's intention in creating this idealized breed of citizenry among the public by implementing the conscription system was clearly indicated in the "Zhengbing ling" (Conscription order), issued on March 1, 1936, the same day as the official beginning of the compulsory conscription law. This conscription order read: "The fundamental way to build the power of the nation is to enlarge and strengthen its military force. The mercenary system separated soldiers from the general masses and consequently prevented national consciousness from developing. . . . Thus [the Nationalist state] encourages people in the whole nation to wake up and perform military duty in compliance with the conscription law in order to recover our nation."[22] The first compulsory conscription law and its related legal and political edicts stressed the crucial role of the military in nation-state-building and conceptually associated military service with citizen status. Having legalized military service as a national duty of citizens, the GMD state intended to bring together the males of vastly different backgrounds in a national bond.

In August 1936—a mere five months after the implementation of the first compulsory law—the Nationalist state made several notable revisions. It introduced the concepts of exemption from military service (*mianyi*), prohibition of military service (*jinyi*), and deferment of military service (*huanyi*). Men who were to be exempted from military service included those who did not meet physical standards, who suffered from incurable diseases, who were only sons in their families, who were assigned by the government with special appointments, or who were students in senior or higher-level schools. Men who were prohibited from performing military service were those who were sentenced to life imprisonment and who were permanently deprived of political rights.[23] Men who could defer military service were civil servants, those who could not fully recover from any disease within months, school teachers, those who were not clear of suspicion for criminal offenses, and those from a family where half of the sons were already active soldiers.[24]

When the Nationalist state issued the first compulsory conscription law in 1933, it did not include the rules on *mianyi, jinyi,* and *huanyi.* The GMD intended to build a unified national identity of disciplined and militarized modern citizens throughout society. Adding these rules immediately after the implementation of the law presents two possible implications. First, the GMD's effort to build a militarized national citizenry that crossed the boundaries of class, profession, family, and educational background had to compromise with reality and the needs of the GMD state's other state-building projects. Giving families' only sons an exemption from active duty recognized that the GMD's state-building agenda was shaped by the deep-seated traditional ideal of filial piety for male descendants. Meanwhile, students in senior or higher-level schools were also exempted from military duty. As the historian Jay Taylor comments, Chiang Kai-shek believed that the "40,000 college graduates in all of China and other educated youth coming on stream were needed to keep the economy and the government functioning."[25] These students' exemptions were necessary for the GMD's state-building efforts to succeed.

Second, the rule forbidding recruitment of those sentenced to life imprisonment or deprived of political rights reveals that the GMD had a clear conception of the ideal archetype of a disciplined citizen-soldier. This rule supported the GMD's advocacy of military service as a prerequisite for a man to achieve full-fledged citizenship. It also allowed the GMD to affirm society's perception that military service was a privilege reserved exclusively for disciplined citizens.

During the Second Sino-Japanese War, the Nationalist state continued its efforts to improve the compulsory conscription system and to exalt the status of the soldier. Another revision that the Nationalist state made in 1939 to the first compulsory conscription law stipulated that soldier recruitment should follow the Principle of the Three Equals (*sanping yuanze*)—namely, impartiality (*pingyun*), equality (*pingjun*), and fairness (*pingdeng*).[26] According to this principle, *bao* and *jia* heads were to take charge of selecting new conscripts, and the responsibility for military service was to rest equitably upon all areas and economic classes of the nation. Local governments were assigned recruit quotas in proportion to the number of able-bodied men within their borders. And then the recruits were to be selected by *baojia* heads drawing lots on a fair basis.[27] By explaining in detail the responsibilities of *bao* and *jia* leaders and the process of recruiting conscripts, the GMD sought to improve the newly established institution—the compulsory conscription law.

The GMD's efforts to improve the conscription law continued throughout the Second Sino-Japanese War. In 1943, the Nationalist state revised the law yet again.[28] The 1933 conscription law had outlined the recruitment system to be a combination of compulsory conscription and mercenary methods, while the 1943 revised law established compulsory conscription as the only military system.[29] According to the second "Zhengbing ling," issued on November 3, military forces were decreed to be the only way to achieve the survival of the nation. It stated, "Without the military force the nation cannot survive, and without the implementation of the compulsory conscription system the military force cannot be reinforced."[30] The revised 1943 conscription law also reinforced the link between military service and citizenship status that the 1933 version cultivated. The 1943 revision added the word "sacred" (*shensheng*) to emphasize that every citizen in the Republic of China had an obligation to perform military service. Adding this word revealed that the Nationalist state wished to exalt the social status of citizens who followed the conscription law and performed military service. The citizen-soldier thus became the model citizen and a national hero to be emulated and respected by society. The 1943 revised conscription law provided that women, too, had an obligation to serve the army, but they were only expected to perform noncombat tasks, such as providing assistance and care to male soldiers. The military remained fundamentally gender-segregated because women were excluded from ground combat.

The 1933 conscription law introduced to society the association between the statuses of citizen and soldier by claiming that performing military service was the duty of every male citizen. Its content was chiefly to provide an explanation of structures, terms, and concepts in the conscription system, such as categories of military service, age requirements, and the drawing and management of new conscripts. In other words, the law was an initial step in establishing the soldier as a citizen who followed state law and fulfilled his obligation of national duty. However, it provided little regulation over the political, economic, and cultural rights and duties that a soldier should have as a citizen.

During the Second Sino-Japanese War, the Nationalist state established a series of edicts regarding the rights of soldiers as national citizens. The "Kangzhan jianguo gangling" (Guidelines for resisting the Japanese and building the state), issued by the GMD on April 1938, was a comprehensive guide to the state's political, economic, military, and diplomatic policies. It stipulated that one of the wartime military policies was "to comfort the injured army officers and soldiers, to arrange a place for the disabled soldiers, and to favorably treat the dependents of the soldiers in order to bolster the military morale and promote national mobilization."[31] The GMD tried to instill the concept that the soldier—as a model citizen and national hero—not only had a duty to perform military service but also enjoyed the individual right to receive benefits from the state.

Soldiers' rights legalized through the GMD's military edicts during the Second Sino-Japanese War covered many aspects of their personal life. For example, according to the "Zhanshi shibing yu jiashu tongxin banfa" (Rules on the correspondence between soldiers and their dependents), issued on May 17, 1939, by the Military Affairs Commission, political workers and copy clerks (*wenshu*) in every army unit were expected to take responsibility for helping soldiers in the unit write letters once a month. Neighborhood leaders were also to mobilize teachers, students, intellectuals, and civil servants to pay regular visits to army barracks and write letters for soldiers.[32] The 1943 revision of the first conscription law expanded rights to active duty soldiers and their families. For example, if an activated soldier had debts that were left unpaid at the time of enlistment, he could delay paying them until after he had completed his active duty. Meanwhile, his spouse and direct family members would enjoy favorable treatment by receiving relief from the local government. In addition, after a soldier retired from the military, he was given favorable consideration when applying for positions in government agencies, social organizations, schools, and factories. The 1943 revised conscription law also had specific rules on the policies of military promotion, honor conferment, and punitive measures. Reservists had the right to return to their original job positions and salaries if they were activated.[33]

The military edicts that the Nationalist government issued during the Second Sino-Japanese War intended not only to uphold the rights of soldiers but also to project the state's influence into every aspect of the soldiers' personal lives. For example, although the Nationalist state supported the soldiers' desires to write letters to their families, it also had specific regulations concerning the letters' content. Wartime correspondence between soldiers and their dependents was limited to discussing the following topics: reporting on safety, happiness, health, promotion, and the honor of the soldiers or their comrades as well as of those coming from the same hometown or lineage; stories of heroic successes in killing and capturing enemies, of loving the nation, of the desire to protect the masses, of being supported and welcomed by civilians; making personal greetings to family members; and discussing family affairs. Soldiers were forbidden to mention certain information, especially the action, direction, and goals of their units, military secrets, and anything that might weaken military morale. The frequency of correspondence was limited to once every month except under special circumstances.[34] When the copy clerks helped the soldiers write family letters, they were encouraged to arouse patriotic sentiments among the family and strengthen their belief in winning the war.[35] The rules concerning soldiers' personal correspondence with their families reveal the Nationalist state's intention of projecting its authority into the soldiers' personal affairs.

The 1943 revised conscription law not only outlined the soldiers' rights but also made clear their duties and limitations upon their rights. First, when the soldiers enlisted and were assigned to units, they were made to swear an oath of loyalty to the Nationalist government. Second, they were taught to guard military and state secrets. Third, they were not allowed to participate in any public gathering or meeting unless expressly permitted by authorities. Fourth, they were required to obtain approval by authorities in order to get married.[36]

The codification of soldiers' rights and duties in military laws formally legalized and institutionalized the link between soldier and citizen status. However, the significance of the rights that the GMD's military laws and edicts gave to the soldiers should not be overvalued. In the first place, the GMD state gave the soldiers rights that benefited their life and family with the intention of bolstering military morale and ensuring continuous sources of new conscripts. These rights were viewed primarily as a means of promoting the state-making interests of the GMD state. In the second place, since the majority of soldiers in the GMD army were illiterate, as the next section will show, there is also the question of just how aware soldiers, especially illiterate soldiers, could have been of their "rights."

The Soldier as the Epitome of Morality

In the almost fifteen years of civil wars in China following the Qing dynasty's collapse in 1911, being a soldier increasingly became associated with warlord army excess. Historian Arthur Waldron notes that warlords' depredations eroded public enthusiasm for the martial spirit advocated by the intellectual elite.[37] The Nationalist state viewed the military as the most effective institution for not only instilling a martial ethos but also promoting moral qualities in the larger society.[38] In Nationalist discourse on the topic of the soldier, the military had a responsibility to train citizen-soldiers into moral models. For the Nationalist authorities, the goal of military education was to cultivate the personality and morality of ordinary soldiers so that they could restore the lost trust in the military among the public. After the soldiers were recruited, the Nationalist government took every chance to cultivate the cadets' moral virtues. The senior-level Nationalist political and military authority Zhu Peide[39] (1887–1937) commented in 1934 that "in every aspect of military training, moral cultivation was the priority."[40] For the Nationalists, a soldier should not only be a model citizen and national hero who performed military service but also should serve as the epitome of morality.

Literacy education was one of the tools that the GMD adopted to cultivate the moral virtues of its soldiers. The Nationalist state decided to operate literacy programs for soldiers mainly because the majority of the Nationalist Army were young and poorly educated. According to a 1929 survey of a brigade in the Third Reorganized Military Zone (*bianqian qu*),[41] 90 percent of its soldiers were younger than thirty, and 73 percent were between the ages of twenty and twenty-five.[42] A 1932 survey showed that the average age of soldiers in a unit of the Nineteenth Route Army was twenty-four.[43] And although the Nationalist state had rules on age when recruiting new soldiers, it did not have specific requirements regarding their educational levels (with the exception of the navy and air forces).[44]

In general, the educational background of Nationalist soldiers was very low. In the first years of the Nanjing decade, the overwhelming impression given of soldiers' educational backgrounds was that the majority of them were illiterate. In 1929, the sociologist Tao Menghe surveyed a unit of 946 soldiers in the Third Reorganized Military Zone, located in Shanxi, and found that only 13 percent of them were able to write letters by themselves. The rest were unable to write and/or read.[45] During the Second Sino-Japanese War, the educational levels of newly recruited soldiers were still very low. After the Battle of South Guangxi that went on in Nanning, Guangxi, in late 1939, a military general tested a group of new conscripts that were drafted from Guizhou. This test found that 97 percent were illiterate and that the remaining 3 percent still were not well-enough educated to serve as copy clerks.

Because of the poor educational background of the young soldiers, the Nationalist army units needed not only to carry out military training for the soldiers but also to conduct citizenship education (*guomin jiaoyu*) for them. Citizenship education included teaching basic knowledge, such as Arabic numerals, measurement units, literacy, and moral virtues. Textbooks were designed not only to teach the soldiers language and Chinese characters but also to further indoctrinate them on the moral spirit of the citizen-soldier that the state expected them to master. Thus, the literacy textbooks edited by the GMD's Military Affairs Commission serve as an excellent window for examining the meaning of the soldier figure that the Nationalist state tried to create.

One of the few available literacy textbooks for soldiers was the 1935 version of *Shibing shizi keben, disance* (Literacy textbooks for soldiers, level three).[46] Lessons in these literacy textbooks were very short, but as a whole present a significant snapshot of the interpretation of the soldier figure by the Nationalist state. The first part of this textbook introduced emblems and symbols of the Nationalist state, including the flags of the Nationalist Party, state, and army, the Nationalist Party anthem, the Three Principles of the People, and the National Day (October 10).[47] The second part sought to cultivate moral virtues in soldiers. These moral virtues included courage, diligence, frugality, solidarity, no fear of death, loving their families, and extending love further to their hometowns, provinces, and the nation.[48] The third part introduced some basic rules and skills that citizen-soldiers should know in order to execute military discipline and deal with the civilian community. For example, the lessons taught the soldiers how to draft important routine documents, such as a request for leave (*qingjia tiao*), a house rental contract, a receipt, a loan note, an invoice, and a lender's note, and also included a primer on how to keep financial accounts.[49]

A review of the literacy textbook for soldiers shows that the GMD state aimed to make the soldiers into politicized, disciplined, and morally cultivated modern citizens. As citizens, the soldiers were expected to develop a nationalistic consciousness and basic knowledge about the nation. They should follow military and civilian regulations; they also should have not only the qualities that were usually considered important for a combatant at the front, such as no fear of death, but also qualities that were not exclusively associated with a warrior. Indeed, given the realities of limited resources, the GMD state considered diligence and frugality just as important as courage and fearlessness. As Lloyd Eastman notes in a study of Nationalist rule, at the start of the Second Sino-Japanese War about 300,000 troops had received German-type training, but only 80,000 of these were fully equipped with German weapons. The remainder of the approximately 1.7 million men in the Nationalist army were, by European and Japanese standards, badly trained and poorly equipped.[50] The poor equipment of the Nationalist army helps explain why values like frugality and diligence were considered by the GMD state to be important qualities for the soldier. The GMD state taught these moral virtues to the soldiers and tried to make them into the exemplification of morality to be emulated by other citizens. Cultivating these moral virtues in the soldiers reveals that the GMD state desired the citizen not only to be disciplined and politicized but also morally cultivated.

The rhetoric that the soldier was not only a disciplined and politicized model citizen but also the incarnation of moral virtue appeared in slogans in other aspects of military education as well. The *Zhanshi lujun jiaoyuling cao'an* (Draft of education decree in the wartime period) that was issued as the textbook for military units and academies by the Branch of Military Training (*Junxun bu*) in September 1944 listed ten maxims for soldiers. These maxims were:

1. To carry forward the Three Principles of the People and defend the nation.
2. To support the Nationalist government and obey the authorities.
3. To respect officers and protect the masses.
4. To be dedicated to military duty and execute the order.
5. To adhere to the discipline and be brave and firm.
6. To foster the sense of solidarity and cooperation.

7. To build senses of responsibility, honor, and shame and uphold martial virtues.
8. To work diligently and practice thrift and simplicity (*jiandan*).
9. To follow etiquette and keep upright demeanor.
10. To cultivate moral character and adhere to the faith.[51]

The literacy lessons in 1935 military textbooks and these ten maxims reveal that during both the Encirclement Campaigns (1931–1934) launched by the GMD government with the goal of destroying the developing CCP army and the Second Sino-Japanese War, the GMD state tried to cultivate many moral qualities into the soldiers and to transform the citizen-soldier into the embodiment of morality.

The GMD's discourse presenting the soldier as a moral model of the national citizen was furthered through its political propaganda from the New Life Movement (*xinshenghuo yundong*). This movement, which was set up in 1934 by Chiang Kai-shek and his wife, Song Meiling (1898–2003), attempted to rejuvenate the Chinese nation and foster moral regeneration in the face of foreign aggression and Communist ideology. Chiang used the Confucian notion of self-cultivation to justify his efforts to regulate the behaviors of citizens and to inculcate the idea that national salvation lay in dictating behavioral norms for individual Chinese. As van de Ven reveals, the movement tried to make Chinese life more military, more productive, and more aesthetically pleasing (the "Three Transformations"—*junshi hua, shengchan hua, meishu hua*).[52]

The New Life Movement attempted to inculcate the public with a basic set of Confucian values and celebrated these so-called native virtues as the principles of national citizens. According to the pamphlet *Junguan de xin-shenghuo* (New life of army officers) written by the Nationalist military authority Zhu Peide, the main intention of launching the New Life Movement was to renew the four traditional virtues—propriety (*li*), righteousness (*yi*), honesty (*lian*), and shame (*chi*)—and to make every aspect of social life comply with these virtues. He viewed the advocating and observing of these virtues as the key to recovering the nation's spirit.[53]

For Nationalist authorities like Zhu, soldiers were the group "who were most suitable to serve as the practitioners of the virtues advocated in the New Life Movement."[54] In Nationalist propaganda, a soldier was the epitome of morality and should play a model role in implementing the New Life Movement. To justify this propaganda, the Nationalists stressed that the virtues celebrated in the New Life Movement applied perfectly to the military. When the previously mentioned military education textbook *Zhanshi lujun jiaoyuling cao'an* spoke of martial virtues, it wrote that the senses of propriety, righteousness, honesty, and shame were the only spirits that the soldiers must possess.[55]

To justify the applicability of these virtues in the military, the Nationalist authorities reinterpreted traditional virtues in their propaganda on the New Life Movement. According to the "Xinshenghuode xinyiyi zhaiyao" (Abstract of new meanings of the new life), a digest of a speech broadcasted on February 18, 1939, to celebrate the fifth anniversary of the New Life Movement, propriety meant decent behavior in peacetime and strict discipline in wartime; righteousness meant justifiable action in peacetime and heroic sacrifice in wartime; honesty meant clear judgment in peacetime and frugality in wartime; shame meant law-abiding consciousness in peacetime and brave fighting in wartime.[56] This abstract of peacetime-wartime corollaries clearly illustrates that the virtues of propriety, righteousness, honesty, and shame, which were advocated in the New Life Movement, also applied to the military. For example, following the conscription law, seeing an army off as it marched to the front, along with comforting and serving the injured soldiers, was evidence of practicing propriety. Fighting bravely against the enemies on the battlefield and volunteering to join the army were both evidence of practicing righteousness.[57] Cherishing government and army resources without any waste was evidence of practicing honesty. Avoiding military service and escaping from battle were evidence of having no sense of shame.[58] By arguing that the virtues advocated in the New Life Movement applied to the military, the GMD state expected the soldier to perform a model role in cultivating moral virtues.

The Nationalist state's expectation for the soldier to be a model citizen and an epitome of morality applied equally to reservists who were not activated or had already retired from active duty. According to the "Lujun zhaoji zanxing guize" (Temporary rules on convening reservists), issued by the Military Affairs Commission on November 2, 1936: "Reservists should pursue their original professions at their neighborhood in peacetime, strictly observe discipline, take physical exercise, love the nation, and serve as the advocator for the common masses. Under the threat of natural disaster or banditry, reservists should be braver and make greater efforts than the common people to defend public interests."[59] By demanding that reservists be braver and make greater efforts to defend public interests, the GMD was sending to society the signal that the soldier, no matter whether he was activated or not, must be a model citizen and set a moral example for the civilians to emulate.

The GMD state's literacy and other military textbooks, political propaganda, and military edicts all showed that it argued for a morally cultivated citizen ideal and constructed the soldier as the archetype for such citizens. As one of the conscription slogans went, "A good citizen is the foundation of a good soldier, and a good soldier is the model of a good citizen" (*liangmin shi liangbing de jichu, liangbing shi liangmin de mofan*).[60] This slogan strengthened the association of the soldier with citizenship status and depicted the soldier as a model citizen. The soldiers were to play a model role for society by developing political consciousness of nationalistic sentiment, fostering loyalty to the Nationalist state, abiding by discipline and regulations, but also in cultivating moral virtues that were advocated by the GMD's propaganda.

The Contrast in Reality

The GMD's military laws, edicts, educational materials, and political propaganda as a whole revealed that the GMD advocated a militarized, politicized, disciplined, and morally cultivated citizen ideal. They also showed that the GMD state tried to elevate the rhetorical status of the soldier as a model citizen who was not only a national hero performing sacred military duties but also an epitome of moral virtues. The soldier thus became a national paragon that the whole of society was to respect and imitate. Exalting the soldier's rhetorical status to such a high level served the GMD's state-building agendas of creating a national citizenry that matched its ideals and achieving nationwide military mobilization.

However, an examination of soldiers' real-life conditions and the implementation of the conscription system on local levels will show that the soldiers did not truly benefit from the rhetorical exaltation of their status; nor did society treat soldiers as model citizens. For that reason, men did not enthusiastically join the army in compliance with the conscription law. In the end, the heroic soldier ideal promoted by the GMD state to achieve military mobilization met resistance from society.

The rough real-life conditions of soldiering also made it hard for society to respect and emulate soldiers as the GMD state had hoped it would.

The soldiers in the Nanjing decade and during the Second Sino-Japanese War were still indisputably of a lower social status, and their lives had not really improved despite the multiple government policies specifically tackling the subject of soldiers' rights. The conscripts were still paid poorly and fed badly, contrasting greatly with the imagery developed by the GMD's rhetoric. There was also a sharp disconnect between the propaganda and the reality of the soldier's status as model citizen, national hero, and epitome of morality.

When the Nanjing regime was first founded, the salaries of ordinary soldiers were decent enough to support a family. Around the year 1927, the monthly salary of a staff sergeant (*shangshi*) was 20 *yuan*, a sergeant (*zhongshi*) earned 16 *yuan*, a corporal (*xiashi*) received 14 *yuan*, a private first class E-3 (*shangdengbing*) earned 12 *yuan*, a private E-2 (*yidengbing*) recieved 10.5 *yuan*, and a private E-1 (*erdengbing*) made 10 *yuan*.[61] Based on the 1958 memoirs of the Nationalist leader Li Zongren, "According to the living standards at that time, a soldier spent about 2 *yuan* on food every month, so an ordinary soldier's monthly salary was enough to support a household of two persons."[62]

After 1932, however, the basic living conditions of soldiers consistently deteriorated. According to the "Kunnan shiqi lujun gebudui zanxing jiyu guize" (Temporary rules on army pay during the difficult period), issued by the Nationalist government in June 1932, the soldiers should have been paid *guonanxiang* ("national-calamity-pay"), which deducted about 25 percent from the original salary.[63] Drawing from a 1935 table on the national-calamity-pay for a cavalry division, after the deduction a staff sergeant was paid 15 *yuan*, a sergeant 12 *yuan*, a corporal 11 *yuan*, a private E-3 8.5 *yuan*, a private E-2 7.5 *yuan*, and a private E-1 7 *yuan*. After the monthly expenditure on food was deducted from the salary, there was little left.[64] Based on a survey of the living conditions of soldiers in a Jiangxi unit done in the middle 1930s, every soldier needed to spend almost 4 *yuan* per month in total to feed himself.[65] With the national-calamity-pay deduction, the lowest-level soldiers holding pay grades of E-1 to E-3 could hardly support themselves, let alone their families. Minister of War Chen Cheng (1898–1965) argued in 1939 that "Inflation makes it difficult for a soldier to afford food expenditures, let alone other life needs. Furthermore, it is now impossible for troops that have been called up to assist their families."[66] The poor pay of the soldiers and their inability to support their families and even themselves prevented the public from viewing them as models. Thus, the GMD's military construction of the soldier as national paragon to be copied by the masses did not generate the intended results.

Not only were soldiers poorly paid, they were also badly fed and frequently suffered from poor health. In the Eighth National Political Council Conference (*Guomin canzhenghui*),[67] held in September 1944, 122 council members presented the "Taolun gaishan guanbing ji gongjiao renyuan daiyu banfa'an" (Proposal of improving the treatment of army officers, soldiers, civil servants, and school teachers). The proposal pointed out that "in recent years with prices increasing dramatically, the soldiers at the front received very low pay and looked famished; many of them were even starved to death."[68] Out of the 14 million that the Nationalists recorded as having been conscripted, about 11 million had either deserted or perished.[69] The loss of new recruits was largely due to lack of medical care, cruel treatment from the conscription guards, and high disease and death rates. Eastman states, "Fearful that the conscripts would try to escape, the guards treated them like prisoners; frequently they roped the recruits together, linking them like pearls on a string."[70] The poor treatment of the soldiers made it unlikely that society would wish to emulate the soldier, the national model in state and party rhetoric, and become disciplined, military-ready, politicized, and morally cultivated citizens.

Poor treatment of Nationalist soldiers was also revealed by many Western correspondents through their eyewitness accounts of wartime China. For example, New York–based *Time-Life* correspondents Theodore H. White and Annalee Jacoby, in their critical reports of Chiang Kai-shek's Nationalist army, detailed the atrocious conditions of soldiers slogging away at the front.[71] Jack Belden, who became a front-line correspondent for the United Press after the war broke out and later for *Time* and *Life* magazines, also remarked on how soldiers were frequently beaten by officers with bamboo rods and treated more like "dumb beasts" than people.[72]

Since the view of much of society was that being a soldier would not bring much life improvement or substantial benefits, the conscription system met resistance. The principles of equality and fairness that theoretically guided the process of drafting new recruits were not fully implemented in reality. Eastman reveals several reasons why the conscription system did not work at the local level. The lottery system for selecting conscripts was seldom employed, and it frequently provoked riots in local neighborhoods where it was used. Many well-off families got their sons into universities so that they would be exempted from the draft. The Nationalist government exercised little control over the *baojia* heads, upon whom it depended to carry out the work of selecting and organizing the draftees. Local authorities changed the population registers, increasing or decreasing recorded ages so that members of their families and their friends' families could avoid the draft.[73] In fact, the persons who were actually in charge of the conscription process at the local level sometimes were not even *baojia* heads. In some counties, local power-holders, such as the *tuhao lieshen* (local bullies and evil gentry) and secret societies controlled all aspects of government. "When the *baojia* heads received a draft call, they would have to convene a meeting of the local power-holders to decide who would, and who would not, be drafted."[74]

Hans van de Ven reveals that not only peasants resisted conscription; urban and rural gentry and landlord families, whom *baojia* heads could not control, also exploited their skills, connections, and money to prevent their children from being drafted. Van de Ven notes, "Large-scale recruitment, sanctioned or not and even if partially fictitious, generated social upheaval, led to widespread evasion, created opportunities for malfeasance, and fueled banditry."[75] Confusion in the implementation of the conscription system confirms that the larger society resisted the GMD's rhetoric elevating soldiers' status.

This chapter mainly focuses on the GMD state's discourse on the soldier figure as forged in its propaganda on the compulsory conscription system, which was initiated in 1933 and improved in the mid-1930s and during the Second Sino-Japanese War. It investigates the Nationalist state's efforts in constructing the soldier as a model citizen who not only performed a sacred military duty in compliance with the conscription law but also cultivated moral virtues as advocated by military education and political propaganda. In this discourse, the soldier took on the mantle of a national hero and paragon that was to be respected and emulated by all of society. This discourse served the GMD's military mobilization goal and its state-building agenda of creating a militarized, politicized, disciplined, and morally cultivated citizenry.

To achieve the goals of state-building and military mobilization, the GMD expanded state institutions, trying to penetrate its regulation deeper into local society. It reintroduced a traditional local control system of mutual surveillance—the *baojia* system—between 1932 and 1934 to implement the conscription system. The first compulsory conscription law, issued in 1933 and implemented in 1936, and its later revisions legalized the linking of soldier and citizen status and reveal the GMD's militarized, politicized, and disciplined citizen ideal. By making clear the soldier's rights and duties as a citizen in military law, the GMD state tried to create disciplined citizens first in the army. By asserting military service to be a sacred duty, the GMD state thereby elevated the status of citizen-soldiers as model citizens and national heroes. Cultivating citizen-soldiers' nationalistic sentiment and moral virtues via literacy textbooks and other military training materials meant that the GMD state expected citizen-soldiers to be politicalized and morally cultivated. The GMD state expected soldiers to be models cultivating the moral virtues it advocated in the New Life Movement starting in 1934. Forging this rhetoric on the soldier figure helped the GMD state propagate the military-ready, disciplined, and morally cultivated citizen ideal to the society.

Much to the GMD government's dismay, the elevation of the rhetorical status of soldiers as national heros and paragons did not lead the greater society to emulate them and actively perform military service. The Nationalist state's delicate efforts in elevating the soldiers' status were not translated into a fundamental improvement of soldiers' living standards. And because of financial shortages, bureaucratic corruption, and poor management, most common soldiers were not able to achieve economic independence for their families or themselves, as was expected by both the state and by society. The view that being a soldier equated to a miserable life was widely held in society, which made the implementation of the conscription system unworkable in reality. The GMD's heroic soldier discourse was resisted by society, which in turn hindered the GMD's state-building agenda.

3

Wartime Soldier Support by Urban Intellectuals and Professionals

Viewed together, the civic education at the Whampoa Military Academy, the military laws, and the political propaganda reveal that the GMD constructed a heroic discourse that celebrated the soldier's status as a model disciplined, politicized, and morally cultivated citizen, someone society should respect and emulate. This discourse became a central component for the GMD's state-building and military mobilization agenda. However, the GMD's attempts to extend this discourse to the national army and society were impeded by resistance from some Whampoa cadets, provincial warlords, and local society, and also by the GMD's own inability to improve the treatment of its soldiers.

After the full-scale war against Japanese invasion broke out in 1937, the depiction of the soldier image was not just a key ideological project in the Nationalists' efforts of state-building; it soon became a central theme that appeared in the writings of the public society. This chapter examines how urban intellectuals and professionals undertook soldier support activities during the Second Sino-Japanese War. It focuses on how the depictions of soldiers through war reportage and documents created by these urban intellectuals and professionals revealed their relations with the state. The chapter demonstrates that these urban forces collaborated with—and also complicated—the GMD's heroic rhetoric of the soldier figure. In this manner, these urban intellectuals and professionals advocated for their political influence as propaganda workers, social mobilizers, and army educators while also voicing their criticism of the GMD government.

The Other Side of the Heroic Soldier Ideal

After the Second Sino-Japanese War broke out in July 1937, many urban intellectuals, such as writers, newspaper editors, rank-and-file college students, and urban professionals, provided services for the army by visiting the front line or by working at local hospitals. Some of them, such as Hu Lanqi (1901–1994) and Xie Bingying (1906–2000), organized the War Service Corps (*zhandi fuwutuan*) and went to the battlefield to report their war experiences and provide assistance to the soldiers. As early as the outbreak of the war in 1937, the GMD state tried to increase its influence among urban intellectuals by drawing writers into its official ranks. Many of these intellectuals who performed their war service were affiliated with the Nationalist government. For example, Guo Moruo (1892–1978) was appointed in 1938 as head of the Third Section of the National Military Council's newly created Political Department in charge of propaganda.[1] As Leo Ou-fan Lee points out, the writers' propaganda activities became "formally sanctioned by the government."[2] The actions of urban intellectuals through war service represented their active participation in the GMD's state-building agendas of army support and national salvation.

The urban intellectuals who organized the War Service Corps identified multiple roles for those involved. Hu Lanqi, who organized the Shanghai Professional Women War Service Corps (*Shanghai laodong funü zhandi fuwutuan*) on October 5, 1937, explained the four main goals of the corps in a 1939 work report: to promote national resistance and mobilize the masses; to detect traitors; to comfort and encourage soldiers, treat and cure the wounded soldiers, and help maintain military discipline; and to investigate the work conditions of civil servants at the local level.[3] The journalist Liu Naifu (1909–1939) summarized in October 1937 the two main tasks of cultural intellectuals during the war. The first was to serve the army by going to the front lines in person, and the second was to mobilize and organize the people by working with urban institutions.[4] The work reports of the War Service Corps organized by these intellectuals and their writings on war service reveal that they performed many roles of war service, such as propaganda work, social mobilization, and war reportage.

Upon visiting the front lines, many notable intellectuals, such as the female writer and soldier Xie Bingying, journalist and social activist Fan Changjiang (1909–1970), artist and social activist Li Mingjian (1916–1979), and Wu Dakun (1916–2007), an economist and the secretary in charge of the propaganda branch of the United Association of Various Circles for National Salvation (*Quanguo gejie jiuguo lianhehui*),[5] as well as many young college students, wrote reports on their war experiences and published them in various wartime newspapers, such as Tianjin's *Dagong-*

bao (*L'Impartial*)[6] and *Jiuwang ribao* (Salvation daily).[7] Their writings met the need of "the young readers of literature and the reading publics who loved to read newspapers and magazines."[8] "Because of their concerns for the war progress, these readers all wanted to see the record and representation of the Second Sino-Japanese War in the literary works with little loss of immediacy."[9] Their writings were later collected in a series of reportages (*baogao wenxue congkan*), such as *Zhandou de suhui* (Quick sketches of the combat) and *Zhandi guilai* (Return from the battlefield), both dating from 1943. These writings provide a window to interpret how soldiers were affected by the social forces of urban intellectuals.

The misery of the soldiers was the direct impression given in the intellectuals' writings about their frontline experiences. The intellectuals often provided detailed descriptions of not only the broken bodies and desperate psychology but also of the poor treatment the wounded soldiers received from the army. Fan Changjiang, in his (1938) reportage "Yi ye zhanchang" (Memories of the wild battlefield), wrote:

> The wounded soldiers gathered themselves into groups of eight or ten. The soldiers with light injuries were walking with many smudges of blood on their bodies. Those with only one leg left hobbled along the road with the aid of a stick. And those who were so injured and unable to move just lay down on the ground and groaned painfully. As these wounded were treated so poorly by the government after fierce combat, it was understandable that they could not bear it. . . .[10]
>
> These wounded soldiers did not have blankets to keep them warm, and their food often had flies inside. For the soldier whose abdomen was injured, the belt was blended with blood and mud into something that you did not have the heart to see. The lower part of the soldiers' bodies was bloodless and powerless, leaning against the wall. The autumn wind mercilessly blew across their faces. It seemed that only the benevolent sunlight could offer them some warmth and comfort, but this comfort was far from satisfying for these soldiers.[11]

Such detailed description focusing on the soldier's body was a theme that often appeared in the writings of the intellectuals. For example, Xie Bingying wrote "Zhanshide shou" (The hand of a soldier) in 1941 to provide a vivid description of a soldier's hand: "I found a soldier's hand covered with blood. Removed from the body, it was lying lonely on the ground. Only the thumb protruded out; the left fingers all slightly bent inward. The hand's skin was very rough. The color of the blood already became dark. Maybe because the hand was cut off not long ago, the carpus still had some bloodstain."[12] The detailed descriptions of the soldiers' damaged bodies had three meanings. First, it complicated the tough and heroic warrior ideal constructed by Chiang Kai-shek and his GMD government. The shock that these descriptions gave to the readers was that although the soldiers fought bravely, they were still ordinary human beings whose fragile bodies were at the mercy of the ravages of war. Second, depicting the soldiers' damaged bodies also expressed the intellectuals' respect for their bravery. Xie Bingying preserved the hand inside a glass bottle filled with embalming fluid, cherishing it as a symbol of this soldier's sacrifice for the resistance war. She wrote: "Do you know how precious this hand is? Its owner has killed countless enemies and performed immortal feats."[13] In her eyes, this hand was a metaphor for the bravery of the soldier. Third, the intellectuals interpreted the damaged body parts not only as metaphors of the soldiers' bravery but also as silent defiance against the poor treatment of the soldiers by the government. In "Yi ye zhanchang," Fan Changjiang remarks, "A dead body had a pair of leathery hands with clenched fists. The fists denoted the soldier's regret of his death before all enemies were killed; they also symbolized that the soldier was so sad and angry with the corruption of the relief agencies."[14] Xie and Fan recognized the bravery of soldiers and gave them their upmost respect, but they also expressed their disappointment with the state's soldier relief work.

The criticism of the inability of the GMD government and the army to provide timely relief for soldiers was explicit in some writings of the intellectuals. Fan wrote: "At the front line, there was no rear service and logistical agencies organized by the army. The treatment and transportation of wounded soldiers relied on the voluntary work of doctors and nurses from local hospitals as well as of local masses."[15] Other reportages revealed that wounded soldiers could not be transferred to hospitals in a timely manner. This pressing issue was characterized by Li Mingjian, who joined the war area service corps organized by Guo Moruo in 1937. In "Zhandi jianying" (The sketch of the battlefield) (1938), she wrote: "The number of the wounded soldiers kept increasing. The soldiers, who had lost either a hand or a leg in the combat, were waiting for the rescuers' vehicles to send them to the hospital. They cried sadly: 'Comrades, come on, please help me. I still need to fight against the enemy.'"[16] In the reportage "Qianxian liang zhouye" (Two days and nights at the front) of 1938, Wu Dakun described how shocked he was when he observed the wounded soldiers: "'Comrades, please do a good deed.' A sad voice attracted my attention. It was a soldier whose two legs were both broken. He was using his hands to crawl on the ground, with blood and mud all over his body. It was apparent that he was begging for help. I was so shocked. We really needed more bearers to send these wounded soldiers to hospital."[17]

The detailed depiction of the soldiers' injured bodies, demoralized demeanor, and pitiable helplessness highlighted the soldiers' bravery and implied the intellectuals' respect and sympathy toward them. The intellectuals' intention of highlighting the soldiers' bravery, however, was not to encourage social masses to emulate the soldier and become militarized citizens, as the GMD had hoped. Instead, the soldiers' bravery, highlighted in their writings, served as a contrast with the misery the soldiers had to bear after they were injured during combat. This contrast allowed intellectuals to criticize the government's maladministration in soldier relief. It also constituted a shocking picture that stimulated great sympathy for soldiers among the readers. Moreover, this contrast allowed the intellectuals to assert the importance of mobilizing the larger society to actively participate in national affairs and public service in regard to soldiers.

In the writings of the intellectuals who visited the front, it was noted that the combination of physical injuries and poor treatment from the GMD led many soldiers to lose hope. The writer Shen Qiyu (1903–1970), who participated in the War Service Corps of Sichuan Native Place Association (*Sichuan tongxianghui zhandi fuwutuan*), wrote in his 1938 reportage "Qianxian guilai ji" (Return from the front line) that "a lot of wounded soldiers did not have any medicine and could not be transported to the hospital because of lack of vehicles. Their psychological condition was unstable, so we need to give them more comfort."[18] Another writer who participated in the war area service corps, Yang Fenjun,[19] wrote in 1938 in "Zai yedi yiyuan" (At the hospital in the front line) describing the pain of the wounded soldiers: "They desperately said to the doctor 'please give me some medicine. I have extreme pain.'"[20]

By depicting the injured bodies and mental desperation of the soldiers, the intellectuals highlighted the contrast with the tough and heroic images of soldiers created in Whampoa's "ten no-fears." The soldiers were described as ordinary human beings who were weak and fragile in war and could hardly bear the pain. Fan Changjiang wrote in "Yi ye zhanchang" that great pain led to violent action by some soldiers. "Because our car could only give a few people a ride, many wounded soldiers were so upset that they beat our car using sticks and guns."[21] Fan further commented that the wounded soldiers' action of beating the car with sticks was "their defiance against the authorities who had the responsibility to provide for them."[22] In Fan's writing, the soldiers were not the epitome of morality, as exalted in the GMD's rhetoric; instead, they performed violent misbehavior when they did not receive adequate relief. Fan, however, did not condemn the misbehavior of the soldiers; instead, using a sympathetic tone, he imputed their misbehavior to the government's inefficient relief work administration, emphasizing the government's inability to mobilize the public to perform soldier support.

Torn bodies, psychological anguish, and uncontrollable tempers became the initial impression that intellectuals had of the soldiers when they visited the front. They also described the soldiers as heroic fighters and as having an optimistic attitude. Fan Changjiang recollected the words told to him by a wounded soldier: "The enemies' artillerymen attacked our village using hundreds of shells. With a fighting will as strong as a mountain, soldiers who were still alive quickly filled up the positions of their dead comrades."[23] In Li Mingjian's reportage "Zhandi jianying," also of 1938, a platoon leader was seriously injured but still kept a clear mind. He told Li that "if he were not seriously injured, he would definitely fight until death."[24] In Wu Dakun's "Qianxian liang zhouye," a battalion vice commander with the surname Li told Wu that "although only four soldiers of his unit survived, they still managed to not only uphold the position but also seize the machine gun from the enemy."[25] Social activist Liu Liangmo (1909–1988) reported in "Zai zhandi yiyuan li" (At the hospital in the front line) one story told by wounded soldiers. They described an injured company commander—whose belly was split open and his intestines falling out—still continuing to fight. His death the next day at the hospital moved other soldiers to tears.[26] Xie Bingying's reportage "Zhanshi de xue ranhongle women de shou" (The soldier's blood tinted our hands red) recorded a story of another company commander with the surname Song. "Song's right hand was broken by the machine gun, but he did not show any expression of pain at all and he still talked with us about the brave fighting spiritedly."[27]

The intellectuals' writings described the soldiers' bravery by stressing not only their fearlessness but also their utter contempt for the enemy. They did this by showing how the soldiers considered the enemies decadent and unwilling to fight. In the reportage "Qianxian liang zhouye" cited earlier, a regimental commander said, "The enemy only relied on the artillery. But none of us was afraid. It seemed that the enemies could hardly hold on and that nobody was willing to keep fighting."[28] In writing this way, these intellectuals echoed the heroic discourse of the soldier figure advocated by Chiang and the GMD government.

By contrasting the miserable treatment of the wounded soldiers with their heroic combat stories, the intellectuals asserted the necessity and urgency of mobilizing social masses to participate in the national affair of war service. These intellectuals often expressed their respect for the wounded soldiers and professed their guilt for not having achieved sufficient social mobilization to support them. Fan Changjiang's commentary in "Yi ye zhanchang" reflected this mind-set: "We are very guilty. These wounded soldiers fought against the enemies so bravely for the nation. They made us proud of them. However, after they were injured, we were unable to relieve their pain by well organizing relief work and immediately sending them to the safe rear. Our government and nationals were really sorry for these brave warriors."[29] Similar comments were echoed in Wu Dakun's "Qianxian liang zhouye": "In the battle of defending Shanghai, our soldiers really did their best for their duty, even though they did not have enough weapons, gas masks and raincoats. On the side of our masses, there were so few bearers in the front line, and the seriously injured soldiers had to bear great pain and crawl on the ground. It was just so unacceptable. I plan to work harder on the social mobilization after I return to the rear."[30] In these writings, the wounded soldiers "symbolized the Chinese people's great spirit of self-sacrifice."[31] This description showed how the intellectuals echoed the GMD state in an attempt to elevate the soldiers' rhetoric status. However, the urban intellectuals expressed their respect for the soldiers not for the purpose of strengthening the militarization of society but with the intent to argue that these brave soldiers who were injured in defending the nation deserved respect from all in society.

The intellectuals complicated the GMD's heroic soldier ideal not only by depicting their fragile bodies as being at the mercy of the brutality of war but also by showing that the soldier was not devoid of emotional expression. This image was described in "Zai shangbing yiyuan" (At the hospital for wounded soldiers), written by a young female college student from Shanghai using the pen name Huizhu and published in the newspaper *Fenghuo* (Flames of war)[32] in November 1937. After Shanghai was attacked by the Japanese in 1937, Huizhu worked at a Red Cross hospital for wounded soldiers as a nurse. This reportage was a record of her daily experience serving the wounded soldiers. The literary critic Tian Zhongji[33] (1907–2002) commented in his book on the history of wartime Chinese literature that this reportage "provided a vivid description of the tragic and intense atmosphere at the wounded soldiers' hospital and the innocent, simple-hearted and kind personality of the wounded soldiers."[34]

In this reportage, the young student Huizhu compares the wounded soldiers to "brave and innocent children who publicly released their emotions, exciting, joyful, sad, anxious, and pessimistic."[35] When they recalled the memories of the combat, they were so excited that their faces turned red.[36] When they were given newspapers to read, their faces showed joy.[37] When they requested the writer to teach them to sing songs, their simple faces showed flames of hope. When the writer accepted their requests, they smiled like children. When the writer sang that wounded soldiers got injured protecting the masses, they were quietly listening with tears in their eyes.[38]

Although Huizhu recognized the soldiers' brave spirit, she also said that it was hard for the soldiers to bear the pain caused by their injuries. The soldier shouted irritably: "I am so full of pain that I want to die. Why don't you give me medicine? Tell the doctor to come here!"[39] When one doctor was cleaning the injury of a wounded soldier, the soldier screamed and sweat covered his head; the doctor comforted him as if he were a child.[40] Some wounded soldiers became so irritable and pessimistic that they refused to cooperate in treatment.[41] Some seriously injured soldiers even wanted to kill themselves, and doctors had to calm them down as if they were talking to children. When their attempts to commit suicide were stopped, they cried like kids.[42] By revealing that the soldiers did not suppress their emotions and that they often cried, Huizhu showed the other side of the tough soldier ideal created by Chiang Kai-shek. By comparing the soldiers to innocent and emotionally charged children, Huizhu argued that the brave soldier was not to be emulated but to be cared for by the greater society.

Women activists in the New Life Movement Committee on Women Guidance (*Xinshenghuo yundong funü zhidao weiyuanhui*) described the wounded soldiers as people who had self-doubt and thus needed to be cared for. Their work report read, "Some of the wounded soldiers were frustrated with their miserable life experiences, and others complained about the government and loathed the national resistance."[43] The women activists maintained that the social masses, regardless of the sex, age, and profession limits, should do whatever they could to show respect and comfort for the wounded soldiers and correct the self-doubt in the soldiers' minds. Women and children could perform consolation and fundraising tasks. Women could mend and wash clothes for the wounded soldiers; children could bring candy and fruit to them; the barber could cut hair for them. These women activists reported that the wounded soldiers were so stimulated and moved by the enthusiasm of the children that they shed tears and requested to return to the battlefield to kill the enemies.[44]

The writings analyzed above emerged from the hands of urban intellectuals who worked in the cultural realm as writers, journalists, economists, social activists, or college students. After the Second Sino-Japanese war broke out, they provided service either by reporting from the front line or by working at hospitals. They recorded what they saw, heard, and experienced, and their writings reflect what they went through. In their comments, they expressed their high respect for wounded soldiers and showed the readers the importance of mobilizing more people to serve them. In the process of performing their army service, they served not only as war recorders and narrators but also as social critics. They viewed their task as being not just to serve the wounded soldiers but also to educate the masses on the significance of participation in social and political affairs.

With these purposes in mind, the intellectuals propagated the brave soldiers' spirit by narrating their heroic combat stories, revealing their contempt for the enemy, and demonstrating great respect for their bravery. In this sense, the intellectuals' writings supplemented the GMD's efforts to promote the heroic soldier image. However, the intellectuals and the GMD had different goals in celebrating the soldiers' heroism. The GMD intended to build a militarized and disciplined citizenry who viewed the soldier as a model, while the intellectuals sought to increase their own political influence as social mobilizers. The writings by the intellectuals complicated the GMD's heroic soldier ideal by revealing the other side of the soldier; the brave and heroic exterior was presented with many characteristics of "normal" human beings: fragility, an uncontrollable temper, hostility in response to injury, and childlike emotionality.

Our Wounded Friends Need Our Education

Urban professionals also participated in war service during the Second Sino-Japanese War. The urban professionals who worked on military welfare advocated larger social participation in serving the soldiers and asserted their political role as social mobilizers and army educators. The social activist Liu Liangmo, who graduated from Shanghai Hujiang University with a major in sociology in 1932, advocated the principle that urban professionals should perform these dual roles. After the Second Sino-Japanese War broke out in 1937, Liu worked at the Emergency Service to Soldiers Program (*Junren fuwubu*) of the Chinese Young Men's Christian Association (YMCA) in Shanghai. The Chinese YMCA, which was founded in 1885 in Tongzhou, Hebei, was guided by the religious beliefs, financial and personnel support, and operational structure of the American YMCA.[45] During the Second Sino-Japanese War, it was the Emergency Service to Soldiers Program that had the largest and longest-running operations of all the Chinese YMCA's sub-branches.[46]

Between 1936 and 1939, Liu Liangmo led a choir and actively organized conferences for the social masses to sing patriotic songs in Shanghai and Zhejiang Province. Liu argued that the undertaking of soldier service should be based on the participation of the social masses, calling on all the nation to serve the soldiers. He believed that the efforts of social communities and individuals to serve the soldiers should be coordinated. Every Saturday he held a forum on how to serve the soldiers in the city of Changsha in Hunan Province, gathering professionals and associations who provided services to wounded soldiers together to exchange work experiences.[47] Liu considered the role of urban professionals who performed soldier service to be "the bridge between the military and the civilian."[48] He aimed not only to mobilize the masses to serve the soldiers but also to educate the soldiers to love the masses in return. He designed a flag of two hands closely clasping each other, one symbolizing the soldier and the other the civilian. He interpreted the meaning of the flag as "the military and the civilian going forward hand in hand."[49] In Liu's eyes, the task of soldier service was to create a military-civilian relationship based on mutual respect and support. Through his arguments, Liu asserted his identities as both a social mobilizer and an army educator.

The Emergency Service to Soldiers Program of the Chinese YMCA was a voluntary organization, and thus it maintained some level of inde-

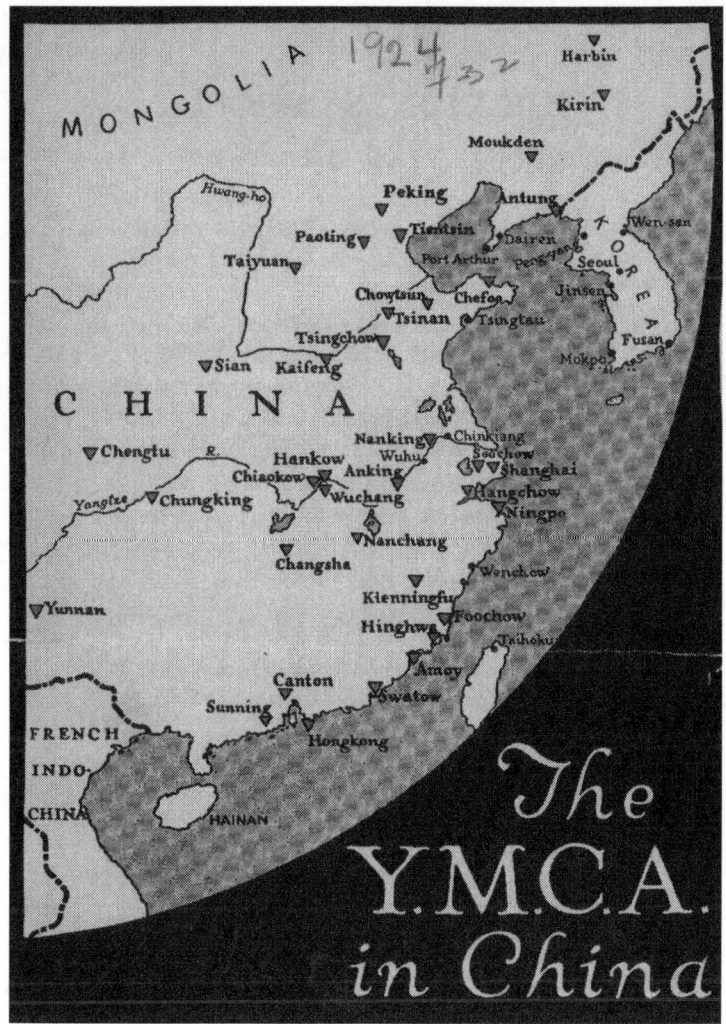

The YMCA in China, 1924.

pendence from the Nationalist state's regulation. Although it received some financial support from the GMD government, most of the funding required for regular operations was obtained through donations by patriotic enthusiasts at home and abroad.[50] In peacetime, the services provided by the program included Bible study, literacy instruction, and intellectual education. Its main purposes were to provide the necessary equipment for the masses to do physical exercise and to guide them to develop healthy ways for entertainment and relaxation. During the Second Sino-Japanese War, however, the program focused on providing services to soldiers, especially those wounded in battle.[51]

The assistance that the Emergency Service to Soldiers Program provided for wounded soldiers was categorized into two types based on availability of equipment. When equipment resources were available, it established soldier clubs, receptions, and medicine-changing stations along railroads and road lines. There, professionals working with the Emergency Service to Soldiers Program offered as many comforts as possible to soldiers: they showed films and slides, performed drama and taught songs, built showers and dining rooms, wrote letters, cut hair, and provided basic medicine and lights.[52] When equipment resources were scarce, the program improvised with activities such as teaching martial arts and gymnastics, chatting, organizing women to wash and mend clothes, teaching patriotic songs, and holding exhibitions to show calligraphy work, paintings, war trophies, and pictures.[53] The program also mobilized and organized the social masses to collect donations and comfort materials, held public speeches, and recruited war service work staff.[54]

The services that the program provided to the wounded soldiers touched a variety of aspects in their daily lives. To the urban professionals working in this program, they believed that the soldiers' education should not be limited to ideological doctrines but should span a breadth of subjects, including martial arts, gymnastics, singing, art, calligraphy, and nationalistic knowledge, etc. Since many of these categories related to the soldiers' daily lives, it was necessary for the program to make further and continuous contact with them. The personnel took it upon themselves not only to take care of the physical health of the wounded soldiers but also to provide them with mental comfort and education. Naturally, this demanded close ties between the military and civilians. To meet this intimate agenda and style, urban professionals working at the program constructed the image of the soldier as an affectionate but lonely human being. Such imagery was depicted in Liu Liangmo's 1938 pamphlet "Zhanshide junren fuwu" (Soldier service in wartime), which provided theoretical guidance on soldier service.

To promote close ties between the military and civilians, Liu Liangmo proposed a discourse that depicted the soldier figure as an affectionate human being who needed emotional support from the civilians and also highly valued sincere friendship with them. He cited experiences in which

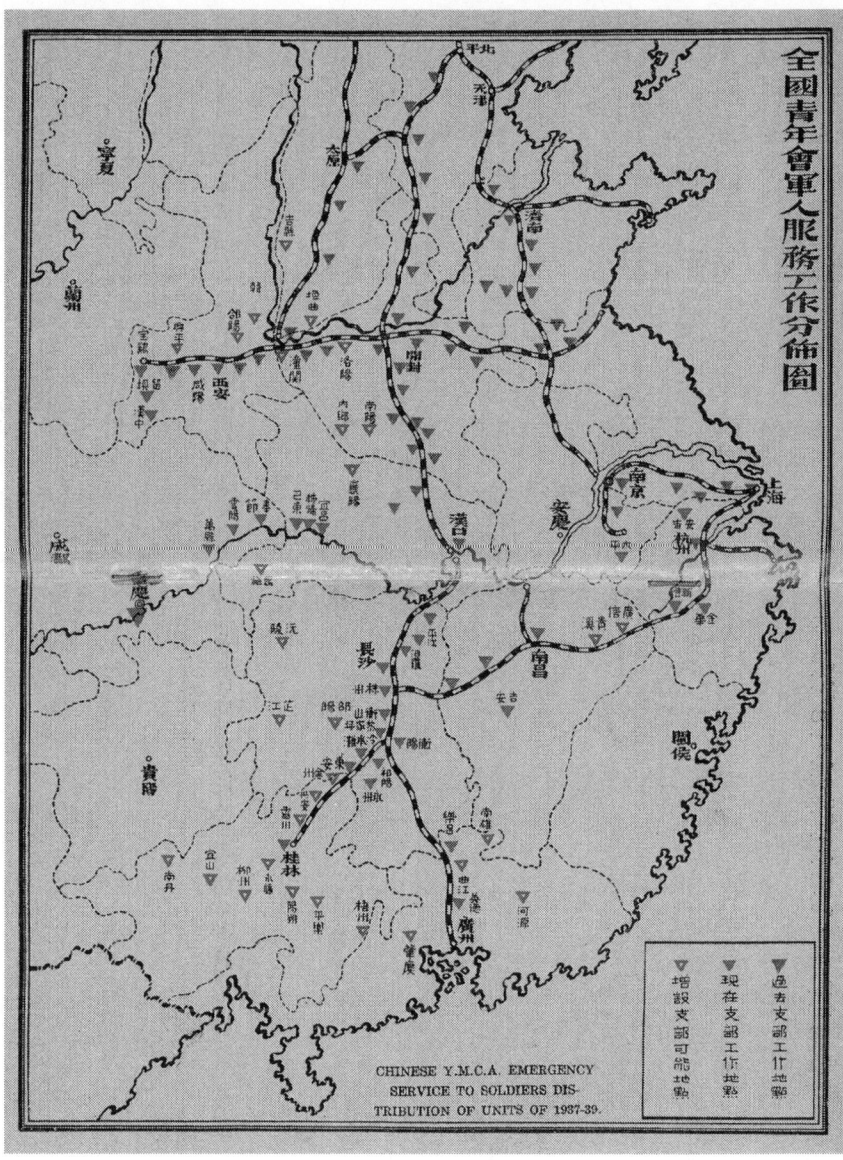

Chinese YMCA Emergency Service to Soldiers—Distribution of Units of 1937–1939. This map appears in the 1939 work report of the National Committee of Chinese YMCA in Shanghai. It shows that by 1939 the Chinese YMCA had provided soldier services in many cities in eastern and central China.

he became good friends with many soldiers while teaching them to sing songs, and said he still maintained contact with many of them.[55] He used his own work experience to show that the key in undertaking soldier service was building sincere friendships and providing the soldiers with sympathy and consolation. He described the soldier as "an innocent and artless child" (*tianzhen lanman de haizi*) and as "the most lonely person who was desperate for the friendship, sympathy and comfort from the masses."[56] To justify his advocating that civilians build friendship with soldiers, Liu pointed out that the wounded soldiers highly respected the well-educated youth who enthusiastically provided service to them.[57] Liu wrote, "The 15,000 wounded soldiers in Changsha not only did not harass the civilians but also generously donated money for the refugees."[58] The image of the soldier in Liu's writing was not only the most loveable person but also the most reasonable and humane. In this sense, Liu embraced ideals similar to those of the GMD government in treating the soldiers as the embodiment of moral virtues.

Liu gave several intimate stories of the relationship between the wounded soldiers and urban professionals to demonstrate that the soldiers were affectionate and friendly people that needed the friendship of the greater society. An illustrative example was of wounded soldiers buying tea and ham and patiently waiting in front of a hospital just to deliver these gifts to the professionals who had served them. Another example was when urban professionals were greeted in the morning at the hospital by the wounded soldiers, who offered to help carry their supplies. During these events, it was common for the wounded soldiers to salute the social workers, showing their respect for the help. These actions led many of the female workers—who were initially afraid of the soldiers and reluctant to serve them—to become more open and willing to think that the soldiers could be friendly.[59] In describing how soldiers' moral qualities made the civilians change their attitudes toward them, Liu collaborated with the GMD government's rhetoric of treating soldiers as the epitome of morality. Similarly, highlighting the soldiers' morality of highly valuing their friendships with the civilians served Liu's agenda of mobilizing the society to participate in soldier service.

Although Liu Liangmo collaborated with the GMD government in depicting soldiers' moral qualities, he also complicated the GMD government's rhetoric of the soldier figure by describing soldiers as common people who were poorly educated and did not fully embrace the nationalistic significance of the resistance war. Liu argued that the soldiers were "armed physically but not mentally" (*tounao hai meiyou wuzhuang qilai*).[60] In Liu's eyes, the historical transformation of the soldier's role from a political tool employed by warlords into a warrior who fought for the national salvation required the assistance of urban professionals for "enriching the soldiers' knowledge and arming their minds" (*chongshi zhishi, wuzhuang tounao*).[61] By stressing the soldiers' need for nationalist knowledge, Liu claimed that the role of urban professionals was not only as social mobilizers but also as army educators.

Liu Liangmo argued that in order to help the soldier "grow into a national warrior in the new era," urban professionals should explain to them the nationalist significance of fighting against the Japanese and teach them to love the civilian masses.[62] The social workers should tell the soldiers to talk gently, to clean the civilians' houses after staying there, to give money to the civilians after buying items from them, to return items they borrowed from the civilians, to pay the civilians after breaking their belongings, and to help out the old, the weak, women, and children.[63] Doing this allowed the soldiers to earn support from the civilian masses. The social workers should also promote the idea that China would definitely win the war and that fighting against the Japanese would save their own lives as well as save the entire nation. The soldiers should also know that they would not win the war without the support from the civilians. With such ideological encouragement, recovered wounded soldiers were expected to return to the front line and continue fighting.[64]

For Liu, the most effective way to serve and educate the wounded soldiers was to establish the Club for Wounded Soldiers (*Shangbing julebu*), where urban professionals could use various methods of entertaining to transform the minds of the soldiers. To justify this idea, Liu described the soldiers as being tired and scared of war. These wounded soldiers "practiced military exercise from day to night and their lives were exhausted and very busy."[65] Liu wrote that the soldiers' injuries during combat made them overwhelmed by the frustration of having been defeated by the enemy and thus unable to see the larger picture of the war, the international situation, and the domestic chaos in Japan; the soldiers' only feeling was being fearful of the war.[66] In this manner, the urban professionals should entertain the soldiers in order to assure and encourage them. The foundation of the club would allow the urban professionals to employ a variety of entertainment techniques, such as singing, drama, and drawing, to educate the soldiers with political knowledge.[67] The purpose of the club was not just to entertain the soldiers; it was also to indoctrinate them into believing that China would certainly win the war, that the soldiers should love and cooperate with the masses, and that the wounded soldiers should return to the battlefield after their recovery.[68] Liu argued that the strategy of education through entertainment was quite popular among the wounded soldiers.[69] By describing the significance of this strategy, Liu tried to assert the political authority of urban professionals in educating soldiers.

Liu's use of entertainment to educate wounded soldiers was also intended to criticize the high-handed measures adopted by many government bureaucrats for the treatment of wounded soldiers. These government authorities believed that shooting some ill-behaved wounded soldiers would prevent more chaos from happening. Liu, however, believed that the coercive measure would only cause more trouble. To him, the proper way to serve the wounded soldiers was to establish a club for them and to educate them with great sympathy. "If all the hospitals in the city had clubs for wounded soldiers, the city would be in peace."[70] After analyzing the benefits of the Club for Wounded Soldiers, Liu encouraged urban professionals to contact the hospital and government authorities, explain the meaning of the club to them, and persuade them to support its establishment.[71] By demonstrating the benefits of using entertainment to educate soldiers, Liu was criticizing the government for its harsh treatment of the men.

The urban professionals at the YMCA, such as Liu Liangmo, echoed the GMD government's discourse of the soldier figure by recognizing soldiers' moral qualities, such as being sincere to friends and valuing friendship. The description of soldiers' moral qualities fit the Emergency Service to Soldiers Program's work style of maintaining constant contact with the soldiers. It also served Liu's goal of mobilizing more people to support the army and serve soldiers. However, Liu complicated the GMD government's heroic discourse by stressing the poor education of soldiers and their educational needs. This allowed Liu to justify the necessity of the urban professionals' political participation as both social mobilizers and army educators. Treating the soldiers as affectionate human beings who had emotional needs, Liu developed the strategy of educating them by incorporating methods of entertainment. He argued that soldier support performed by the professionals at the Chinese YMCA was more effective than the work performed by government agencies. In doing so, he voiced criticism of the GMD government's maladministration of soldiers and proposed the alternate authority of urban professionals at the Chinese YMCA as educators of the military.

Honorable Soldiers Need Vocational Rehabilitation

The Emergency Service to Soldiers Program of the Chinese YMCA, as analyzed above, defined the tasks of army education as strengthening their courage, powering their belief in China's victory, and encouraging them to cooperate with the masses. However, they were not the only group of urban professionals who asserted their role as army educators. Vocational training specialists also claimed their role in impacting the education of the military. They believed that equipping the men with professional skills would be beneficial to the soldiers' long-term career development. Therefore, vocational training was an essential educational regimen for soldiers, especially the wounded.

The vocational training specialists promoted to the GMD government the importance of providing vocational training for wounded soldiers. Unlike the Chinese YMCA professionals, who encouraged the wounded soldiers to return to the battlefield after recovery, vocational training specialists stressed that it was not feasible for every wounded soldier to return to the army and resume fighting. The GMD government expected wounded soldiers, who had garnered substantial fighting experience, to return to the battlefield and improve the combat capability of the army. Chiang Kai-shek's son Chiang Ching-kuo (1910–1988) published a speech to wounded soldiers in celebration of the new year of 1941 in the magazine *Shangbing zhi you* (Friends of the wounded soldiers).[72] In this speech, Chiang stated, "A soldier who has recovered from the injury is as competent as ten new conscripts. This is how valuable our wounded soldiers are."[73] He expected wounded soldiers to return to the fight.

Urban professionals working on vocational education did not dissent from the GMD government's view. The education professional Chen Junming,[74] who worked at an administration bureau for wounded soldiers in Hengyang, Hunan Province, admitted that the soldiers who returned to combat after recovery could immediately serve as a cadre to lead new conscripts.[75] However, Chen pointed out that it was difficult for wounded soldiers, especially army officers, to return to the battlefield. Their original military units usually refused to have them back. While the soldiers or officers were hospitalized, their units found someone else to fill the position.

The replacements became familiar enough with the unit, and the army found it inconvenient to force them to leave. Furthermore, the Nationalist army ordered that all wounded army officers who chose to return to the front line should be promoted. As a result, having a wounded army officer return to his original unit would increase the unit's expenditure. Chen also pointed out that although the majority of wounded soldiers wished to be assigned to their original units, they were uncertain how frequently such wishes were granted. As a result, many wounded soldiers refused to return to the front line.[76]

Chen Junming did not oppose the GMD government's expectation that wounded soldiers should return to the battlefield and keep fighting. However, he believed that such an expectation was not practical. Chen pointed out that providing vocational training and teaching wounded soldiers the skills required for civilian professions were important to the army. The person who first proposed the idea of vocational rehabilitation for wounded and disabled soldiers was the social activist Duan Shengwu (1897-1944). Born in Hebei, Duan joined the army in 1911 and served as a division commander under warlord Sun Chuanfang (1885-1935) during the early years of the Republican era. After Sun was defeated in the Northern Expedition campaign, Duan left the army and worked as a social activist. He mobilized the refugees in Hebei to migrate to the northern places like Baotou and to farm the land. He encouraged the building of a new community called Hebei New Village (Hebei *xincun*), and for those who migrated to establish associations in an attempt to create self-governance.[77]

In 1938, the Political Branch (*Zhengzhibu*) was founded within the Logistics Department of the Military Affairs Committee in Wuhan, Hubei Province, with Duan appointed director. Duan commented: "The treatment of the wounded and disabled soldiers is crucial for maintaining the army morale and stabilizing the society. I feel very satisfied to have the chance of serving them."[78] He used the name "honorable soldiers" (*rongyu junren*) to refer to those who were wounded and disabled during the war. Duan established a committee in charge of educating honorable soldiers and invited influential social figures to design the education plan. He argued that the government should cooperate with urban professionals in providing education and training for these honorable soldiers.

The efforts to provide vocational rehabilitation for soldiers initially advocated by social activists like Duan were soon institutionalized by the GMD government. The Association of Vocational Coordination for the

Honorable Veterans (*Rongyu junren zhiye xiedaohui*) was established on May 12, 1940, at Chongqing. It was a professional association affiliated with the GMD government devoted to promoting the career development of honorable soldiers. Chiang Kai-shek served as the honorary president of the association. H. H. Kung (1881–1967), who served as the vice premier of the Executive Yuan of the GMD government from 1935 to 1945, and Chiang Kai-shek's wife, Madame Song Meiling (1898–2003), served as honorary vice presidents, and Minister of National Defense He Yingqin served as the president.[79] Although high-level GMD authorities were the leaders of this association, it was not an institution totally controlled by the government. The association membership involved a combination of government-affiliated agencies and social forces. Half of the members came from government-sponsored administrative and educational agencies dealing with honorable soldiers, and the other half originated among academic communities and social activists. This membership composition allowed the association not only to manage necessary support from government authorities but also to mobilize larger social forces to research and work on vocational rehabilitation for honorable soldiers.[80] Although the association included active members from government agencies, government subsidies constituted only a small part of its funding; most of the finances required for regular operation came from membership dues and from donations.[81] Social forces who worked on vocational rehabilitation for honorable soldiers did not resist the GMD's efforts to champion their work within a national campaign coordinated by a government-sponsored association. However, the membership composition and the funding sources allowed the association to maintain some independence without challenging the GMD government's authority.

The association established vocational rehabilitation institutes in Chongqing in October 1940 to teach honorable soldiers career skills. The institutes opened a variety of classes, including official document writing, handicrafts, escorting goods, guarding storehouses, receptionist and registrar services, spinning and weaving, carving, political work, accounting, printing, and making various good (i.e., umbrellas, towels, cigarettes, shoes, soap bars, and canework).[82] These classes aimed to provide professional training so that honorable soldiers had the necessary skills to pursue a variety of careers after the war. Within seven years, the institute had opened forty-nine classes, in which over a thousand honorable soldiers participated.[83]

The most well-known professional who studied principles and techniques of vocational education for honorable soldiers was Yu Zhaoming.[84] He had studied vocational education and psychology in the United States in the 1930s and now ran the Yunnan office of the Society for Chinese Vocational Education (*Zhonghua zhiye jiaoyushe*) in the southwestern province of Yunnan during the Second Sino-Japanese War. Upon the request of the Ministry of Society (*Shehuiju*) of the GMD government,[85] he wrote the book *Rongyu junren zhi zhiye zaizao* (The vocational rehabilitation for honorable soldiers) in 1942.[86]

Yu's primary goal in the book was to develop vocational rehabilitation work for honorable soldiers by constructing a large-scale movement where all social members could participate. He viewed the undertaking of vocational rehabilitation for honorable soldiers as a duty for all members of society rather than explicitly the government.[87] According to Yu, the GMD government should not only provide financial, administrative, and legal support for the vocational rehabilitation programs but also should closely collaborate with social communities, such as schools, factories, and farms in the countryside.[88] Yu urged the GMD government to support and cooperate with social forces in this undertaking instead of simply focusing on regulating the action of social forces.

In order to justify the importance of vocational rehabilitation as a necessity requiring soldier cooperation, Yu argued that the soldier was the embodiment of many moral virtues, echoing the GMD government's rhetoric of the soldier figure. Yu stressed that the soldiers—most of whom came from peasant families—possessed moral qualities that were essential for civilian professions, such as conscientiousness, obedience, punctuality, and endurance.[89] These moral virtues would help honorable soldiers manage a smooth transition to the civilian world after they received vocational training.

Although urban professionals echoed the GMD government's rhetoric of elevating the soldier's status as being the epitome of morality, they also pointed out that soldiers were poorly educated and lacked the knowledge and skills required for career-building. To advance this idea, Yu conducted a survey in 1941 on the educational and work backgrounds of 5,869 honorable soldiers. According to the survey, 60 to 70 percent of the soldiers had never received any vocational training, nor acquired any non-military work experience. Only 0.051 percent had graduated from college, while 0.818 percent had graduated from high school, 2.13 percent had graduated from middle school, 41.54 percent had graduated from public elementary school, 4.379 percent had graduated from private elementary school (*sishu*), and 51.082 percent were illiterate.[90] In another survey done in the last year of the Second Sino-Japanese War, Yu compared the educational background between honorable soldiers and workers and merchants (*gongshang renyuan*). It showed that 28 percent of workers and merchants were graduates of elementary schools and 23 percent were graduates of middle schools. In contrast, 30 percent of the disabled soldiers were graduates of elementary schools, but less than 1 percent of them were graduates of middle schools. The lowest educational level of workers and merchants was two years of study at elementary schools, but over half of the honorable soldiers were illiterate.[91] Yu's surveys complicated the GMD's rhetoric of the heroic soldier ideal by stressing the poor education of the soldiers.

Although urban professionals did not oppose the emphasis on the soldier figure's being the epitome of morality, they did point out that the depiction of honorable soldiers as a moral model was ambiguous. If the wounded and disabled soldiers did not receive vocational education from the professionals, they might become a potential source of social disorder. In a 1939 report Chen Junming wrote: "The wounded soldiers had strong regionalism notions. They organized themselves based on their hometowns and established a mysterious and firm network barrier. Sometimes they had serious and even violent arguments with each other over trivial business, but sometimes they gave each other support and care. Therefore, the professionals who served them needed to give them proper guidance."[92] Honorable soldiers were sometimes reported as jobless vagrants who seriously threatened social stability. A rehabilitation house (*jiaoyangyuan*) in Nanxi County, Sichuan Province, wrote a report to the provincial government on June 19, 1943, saying: "The wounded and disabled soldiers created social trouble when we banned smoking tobacco and taking drugs. Yesterday afternoon, they gathered together to attack the county government and beat the police in the street. Their violent misbehavior caused a strike among merchants and students."[93] In the eyes of urban professionals, honorable soldiers were not always a moral model for society; they were instead a potential source of social chaos. Therefore, providing honorable soldiers with vocational rehabilitation and enabling them to support themselves independently and produce for the nation were necessary in order to maintain the social order.[94] By stressing the ambiguity of the soldier's role as moral model, Yu demonstrated the necessity of urban professionals as army educators.

Another way Yu complicated the GMD's rhetoric of the soldier figure as the epitome of morality was to argue that the soldier should expel
any sense of superiority over civilians. According to Yu, the soldier should
try to adjust himself to the civilian community by accepting the reedu-
cation from urban professionals so that he could independently support
himself. Yu argued that honorable soldiers should love the masses as their
brothers and develop a harmonious relationship with them (*dacheng yipian*) by working and living together. In Yu's eyes, there was not a sharp
distinction between honorable soldiers and civilian masses; everyone was
a citizen who worked for the nation.[95] By advocating this stance, Yu tried
to enlist soldiers' cooperation in receiving vocational training from urban
professionals.

This chapter has examined soldier support activities performed by urban
intellectuals and professionals. It focuses on how these social forces constructed the discourse of the soldier figure to meet their work agendas.
It does not aim to evaluate the effectiveness of army service and social
mobilization performed by these urban publics. Instead, it attempts to
demonstrate that these urban intellectuals and professionals asserted their
political influence as social mobilizers and army educators by both echoing and complicating the GMD government's heroic rhetoric of the soldier
figure.

In many respects, urban intellectuals and professionals incorporated
rhetoric similar to that of the GMD government. The service they performed supported the GMD army's resistance against the Japanese and the
GMD government's wartime state-building agenda of supporting the army.
Some of the intellectuals and professionals had government-affiliated positions, such as Duan Shengwu. The educational specialist Yu Zhaoming was
invited by the GMD government to write the book on vocational education for honorable soldiers. Associations of urban professionals, such as
the Soldier Service Branch of the YMCA and the Vocational Coordination
Association for the Disabled Veterans, received financial support from the
government as well. These urban intellectuals and professionals wrote articles and books to encourage larger social masses to participate in the army
service work, which also provided the theoretical guidance and experience
for the GMD government's work on wounded soldier relief and administration. The urban intellectuals celebrated the bravery of the soldiers and
expressed their respect for them; they also acknowledged the moral quali-

ties possessed by the soldiers. Their celebration matched the GMD's rhetorical elevation of the soldiers' status as model citizens and national heroes.

Although urban intellectuals echoed the GMD government's discourse of the soldier figure in many ways, they also complicated this discourse and thereby obtained some autonomous space. They did so by voicing criticism of the government and asserting their role as army educators. The intellectuals—who identified themselves as war reporters, propaganda workers, and social mobilizers—created a heroic image of the soldier that was different from the GMD's heroic soldier ideal. Soldiers in their depictions fought bravely as heroes, but their bodies were at the mercy of the ravages of war. They often lost their temper and were prone to violent misbehavior; they did not restrain their emotion as Chiang expected in his speeches at Whampoa; they were affectionate and sometimes emotionally charged. The urban professionals—who identified themselves as both social mobilizers and army educators—drew from their work experience at the Emergency Service to Soldiers Program of the Chinese YMCA and stressed that the soldiers, especially wounded soldiers, had urgent needs for emotional support. They were poorly educated and needed to be taught the nationalist rhetoric and the importance of cooperating with the social masses. For the professionals working on vocational rehabilitation programs for wounded and disabled soldiers, honorable soldiers' role as moral models was ambiguous, as they could be a source of potential social chaos if they did not receive vocational training from the professionals. Soldiers also lacked necessary skills that would enable them to manage a smooth transition to the civilian world. Therefore, urban professionals believed that honorable soldiers should not have any sense of superiority over civilians, but instead should accept vocational training.

The discourses of the soldier figure constructed by these intellectuals and professionals were shaped depending on the political roles they identified for themselves, be it as war reporters, propaganda workers, social mobilizers, or army educators. They echoed and complicated the GMD government's rhetoric of the soldier figure, which revealed that their alliance with the GMD government was ambiguous.

4

Creating Gendered Images of the Soldier Figure in Literary Works

The previous chapter has shown that urban intellectuals and professionals complicated the tough and heroic image of the soldier figure originally created in the Whampoa Military Academy's civic education. This chapter adopts a literary perspective and examines how a particular group of cultural forces—writers, with direct army or combat experiences—shaped their own image of soldier heroes in literature. The topics of literary images of Chinese soldiers and Chinese war literature have been largely neglected in traditional scholarship on Chinese military history and the Second Sino-Japanese War, which focuses instead on the political, diplomatic, and military perspectives. In particular, this chapter provides a close reading of three literary works. The first is *Bayuede xiangcun* (Village in August), a work of fiction published in 1935 by Xiao Jun (1907–1988), who was a Manchurian refugee writer in Shanghai who experienced the war personally while in the warlord's army. The second is from a former Nationalist army soldier-turned-writer, Qiu Dongping (1910–1941), who wrote battlefield reportages (*zhandi baogao wenxue*) that were published in the Shanghai journal *Qiyue* (July) in the early years of the Second Sino-Japanese War. The third is the autobiography of Xie Bingying (1906–2000), a woman writer and one of the first female soldiers to study at Whampoa Military Academy. The gendered images of soldier heroes in these literary works, shaped by the writers' personal backgrounds, either questioned or collaborated with the GMD's political discourse of soldier heroics. In doing so, these literary images asserted their own goals, which were to promote the influence of literary intellectuals as social critics, to achieve personal independence, and to advocate for women's political participation.

Grassroots Soldier Heroes in *Bayuede xiangcun*

Xiao Jun, a soldier-turned-writer, was born in 1907 in Liaoning, Manchuria. Beginning in his childhood, Xiao developed a reputation for having a heroic personality and determined spirit, which together drove him to pursue a military career and later to use the pen to attack social crises and foreign aggression. He had experienced severe trauma as early as seven months old, when his mother committed suicide due to persistent physical abuse by his father. The severe discipline used by his father cultivated Xiao's tough personality. When he recalled his childhood years, he commented, "A soul of revenge was growing from weak to tougher, from faint to brighter. It had been cultivated in my blood, building up little by little."[1] At ten, he suddenly realized, "I understood that since I was a child without my mother, I should be stronger and earn everything by fighting instead of begging for sympathy."[2] As he wrote in his memoir, he continued to dwell upon this topic: "The belief that fighting for everything had developed in my young head. I would conquer everything using my life as the last bullet and never submit."[3] Xiao's toughness was also shaped by martial arts folktales, such as *Yangjia jiang* (Military generals in the family Yang), which were often told by his grandmother.[4] With this childhood experience, Xiao Jun sprouted the ideal of individual heroism, which stimulated the idea of joining the army as he grew up.

In 1925, at the age of eighteen, Xiao Jun joined a unit of the Northeast Army in Jilin, led by warlord Zhang Zuolin (1875–1928), and was promoted to clerical assistant soon after. Xiao quickly realized during this period that the heroic army life, which he used to admire, was in reality full of brutality and darkness. He detested the strict hierarchy and ugly violence committed by his unit, and eventually he traveled to Shenyang to attend the Northeast Army Military Academy (*Dongbei lujun jiangwu tang*) in 1928. At this military academy controlled by the warlord army, he was shocked when he witnessed other cadets kicking the head of a dead soldier like a ball. His first publication, the short essay "Nuo" (Cowardice),[5] repudiated this ugly and inhumane side of the warlord army, but his criticism offended the academy. In 1930 he was abruptly dismissed just prior to graduation, in part because of his published critique, and also because he protected his fellow classmates from violence by one of the military instructors. It was then that he joined another army unit in Shenyang and served as a military instructor. After the Mukden Incident in 1931,[6] when Japanese troops seized the Manchurian city of Mukden (now Shenyang) and occupied all of Manchuria (now northeast China) within months, Xiao Jun requested permission to launch a guerrilla war against the Japanese. His request was rejected, and he resorted to writing as a way to participate in the resistance war against the Japanese.

Xiao Jun's 1935 fictional work *Bayuede xiangcun* was an extension of his fighting spirit into the literary field. This fictional piece details combat stories about a small guerrilla unit consisting of only nine soldiers who organize a Communist-led self-defense corps in a Manchurian village. The narrative concentrates on the lives of the soldiers and on their skirmishes against the Japanese. The story is a combination of true combat stories concerning a local guerrilla unit (taken from Xiao's personal army experiences) and his own artistic creations.[7] According to Xiao, the motivation to write this book was "to pursue national independence and people's liberation."[8] This fiction de-idealized the heroic discourse of the soldier figure as constructed by Chiang Kai-shek and the GMD government in many ways. The fighting heroes in this fiction come from the social grassroots and are guerrilla soldiers under the CCP banner; they are also chivalrous, fearless, and motivated to fight for personal pursuits. Edgar Snow writes in his introduction to Xiao's novel: "Here is no black-and-white tale of villainy and bestiality on one side and saintly perfection on the other, but a realistic report written by a soldier, filled with enthusiasm for the whole story of . . . ordinary mortals."[9]

In this work of fiction, national heroes who bravely defended the land are not the highly disciplined, politicized, and morally cultivated soldiers that the GMD government and the Whampoa Military Academy tried to produce. Instead, the images of war heroes in this fiction are of common people from the lower social strata of society who not only never received formal military education but also are ideologically incongruent to the Nationalist mentality. The characters depicted included a former shoemender (*fengxie jiang*), a bandit, a warlord soldier, and a student. They do not even receive full names. Their names are simply based on their nicknames, their gender, or the order of birth from within their family, such as Third Zhao (Zhao San), Erliban (Two Miles and A Half), Green Hill Li (Li Qingshan), Old Woman Wang (Wang Po), Metal Eagle (Tie Ying), Little Red Face (Xiao Honglian), Third Brother Li (Li Sandi), Tall Liu (Liu Dagezi), Old Lump Tang (Tang Lao Geda), Seventh Sister Li (Li Qisao), Lark Bird (Bailing Niao), and Freckled Face (Mazi Lian).

These guerrilla soldiers fight not because they are cultivated with nationalistic sentiment or contain a strong belief in the Three Principles of the People. Rather, they join the war for their own unique personal motivations. Little Red Face fights so that one day he can freely plow the farmland with a tobacco pipe and kill everyone that has bullied him and forcibly occupied his land.[10] He also hopes that his wife will not starve and his children can study at a school instead of collecting coal cinders near the railways.[11] For other soldiers, the admiration from other men and their lovers motivate them to fight. For instance, when a factory worker who admires Old Lump Tang's weapon gives Tang a salute, Tang proudly feels that he is "pursuing a glorious undertaking."[12] Tang often shows off to his lover, Seventh Sister Li—whom he tries his best to protect—by boasting that he possesses a gun and is a member of the revolutionary army.[13]

The soldiers' fighting motivations are also gendered. Although male guerrilla soldiers fight the war to protect their families and lovers, Seventh Sister Li feels as if she has no recourse but to join the guerrilla unit after her baby is killed and she is raped by the Japanese. After her lover Tang is shot by the Japanese, she strengthens her resolve even further, determined to exact revenge for him by continuing to fight. Xiao's stress on the personal suffering of the guerrilla soldiers was an attack on the darkness of rural Chinese society and the Japanese invasion.

The guerrilla soldiers in the fictional work do not interact with each other by following the principle of fraternity advocated in the motto of the Whampoa Military Academy—"Qin, Ai, Jing, Cheng" (intimacy, fraternity, dexterity, sincerity). Instead, their relations are largely dependent on their own personal whims and interests, as the guerrilla soldiers compliment their favorite comrades and often argue with those that they dislike. Their relations are not based on a rigid hierarchy but on a fairly casual relationship structure. The soldier Xiao Ming says: "None of us is an officer. We are brothers instead. We have to bear any difficulty and hardship in order to avenge our dead brothers."[14] When these guerrilla soldiers are marching, they chat with each other on a variety of topics, including women, commanders, and battlefield stories. When they argue, they talk using dirty and vulgar words without getting upset.[15] The unit leader does not criticize their arguments; instead, he believes that small arguments among soldiers will help them forget the fatigue, fear, and nervousness endured during combat.[16]

Although these peasants-turned-guerrilla soldiers were never formally trained and disciplined by a military education, they are eager to prove that they are brave, hard-nosed, and hold no fear of or sympathy for the enemy.[17] The peasant-born Metal Eagle, leader of the guerrilla unit, had been a bandit before he joined. Metal Eagle is stern to his subordinates and never shows leniency when he fights. His nickname is purposely used to symbolize his ferocity and agility.[18] Soldier Xiao Ming, who begins as a student, also believes that "a revolutionary army soldier is absolutely not allowed to vacillate or feel depressed."[19] These peasants never attended a military academy nor received any discipline or education from the state; however, they still develop the mental fortitude to fight bravely.

These grassroots-based soldiers have the resolve to fight bravely while relying on each other for survival. Tall Liu wants to be accompanied by other soldiers so that he can push himself to be brave and not lose face in front of others.[20] His logic of mutual reliance is different from the principle of *Lianzuofa* advocated by the Nationalist army. *Lianzuofa* forced soldiers to fight bravely by requiring them to supervise each other and ensure that none would flee from the battle. This writing presents a differing opinion: that the quality of bravery to the point of fearlessness is not built through military discipline. It espouses the idea that a soldier does not have to receive any military and political discipline to be a fearless hero. These guerrilla soldiers refuse to retreat not because they are hardened by the discipline of military training but out of fear of being shamed in the eyes of others. This work of fiction presents an image of war heroes who are not the idealized citizen-soldier of the Nationalists but are common villagers who give a concerted effort to fight the enemy.

Continuing with the intention of projecting an image of national heroes as ordinary human beings, Xiao describes how the guerrilla soldiers also have natural sexual desires, as they are conscious of their sexual attractiveness to women. When Tall Liu discusses the concubine of a company commander, Third Brother Li says: "If you had longer neck and thigh, smaller head and better-looking face, you would be handsome enough to make her run away with you."[21] When describing Old Lump Tang's lover, Seventh Sister Li, to Metal Eagle, one soldier focuses on her sexual body: "She is hot with big breasts and thick lips."[22] Tall Liu stares at the face of a nearby female soldier when writing down a commander's order, feeling that her voice matters even more than the order.[23] When Metal Eagle hears from other soldiers that Seventh Sister Li will not even talk to a man she dislikes,[24] he feels that his body has been "hit by an instinctive force and enveloped by jealousy."[25] These peasant soldiers' expressions of their sexual desires de-idealize the Nationalist leader Chiang Kai-shek's heroic ideal of soldiers. Chiang tried to discipline the soldiers' emotions, limiting them to expressing feelings only for the nation, their family, their parents, and army comrades. Chiang's ideal of a model soldier was an asexual hero; however, the heroes in Xiao Jun's fiction are ordinary villagers who harbor sexual instincts and do not restrain from expressing them openly.

By creating the image of national heroes as common peasants, Xiao expresses his concerns not only about the misery of the social masses but also about the promise of Chinese nationalist resistance. As Lu Xun writes in the preface to this work of fiction: "It is serious and tense. The emotions of the author, the lost skies, earth, the suffering people, and even the deserted grass, *gaoliang* (sorghum), frogs, crickets, and mosquitos—all are muddled together, spreading in gory-red color before the very eyes of the reader, revealing a part and whole of China, contemporary and future of China, a dead-end road and an open road for China."[26]

The guerrilla soldiers in this writing draw their fighting motivation from their instinctive yearning for a peaceful and prosperous family life, as well as their willingness to protect their lovers and exact revenge for those lost. They do not concern themselves with nationalistic ideology or devotion to the Three Principles of the People. Furthermore, they are part of the forces that are led by the Communists, not the Nationalists. They also do not hide their sexual desires. Although they are not as disciplined and politicized as the GMD state expected from its citizen-soldiers, they are still portrayed by Xiao Jun as war heroes that bravely defend their village. As a contemporary literary critic commented, this fiction reveals the social basis and major force of the Chinese revolution.[27] For Xiao, it is the soldiers formed from these social grassroots that constitute the vanguard of Chinese national resistance and the hope of Chinese recovery.

As Leo Ou-fan Lee comments, "The authenticity of feeling—the emotions of Xiao Jun from his immediate experiences—accounted for the fiction's instant popularity."[28] The vivid descriptions of suffering villager soldiers and their subsequent strong motivations to protect or seek revenge in the name of their families were shaped by Xiao's philosophy of individual heroism, molded from his direct observations of the misery experienced by common soldiers in the warlord army. Xiao actively participated in the national resistance campaign as a social critic and fiction writer by revealing the brutality of the enemy and, by extension, the entire war. He also pointed out that grassroots efforts in rural China were a potential social force for national salvation. If widely mobilized and well organized, this force would contribute heavily to national resistance. Utilizing literary writing to reveal the war's brutality and the common peoples' misery, Xiao de-idealized the Nationalist discourse of a politicized and disciplined soldier figure in many ways. His realistic portrayals allowed him to assert the role of literary writers as social critics.

War Trauma in Qiu Dongping's Battlefield Reportage

Xiao Jun's 1935 work *Bayuede xiangcun* depicts peasants-turned-guerrillas fighting in rural China. Images of other types of soldiers also appear in the literary genre of battlefield reportages. The outbreak of the full-scale war against Japan catapulted battlefield reportage, a literary genre often used by Chinese writers to describe war progress and soldiers' combat experiences, into an unprecedented position of prominence. According to commentary by the literary critic Yi Qun (1911–1966), during the war all literary journals devoted 70 to 80 percent of their space to reportage, and 80 to 90 percent of authors, including fiction writers, poets, essayists, and literary critics, wrote at least a few reportage pieces.[29]

Many combat reportages, including Qiu Dongping's battlefield stories, were published in *Qiyue,* the first major wartime literary journal.[30] Hu Feng (1902–1985) founded this weekly journal in Shanghai on September 11, 1937, in response to the breakout of the Second Sino-Japanese War.[31] Its small circulation was funded primarily by donations from Hu Feng's friends.[32] However, a strong desire for independence motivated this small journal from its beginning. As the literature scholar Yunzhong Shu comments, *Qiyue* differed significantly in subject matter and perspective from "official journals," which were "associated in spirit with the government propaganda apparatus" and "intended to sing praises of the Chinese nation for its heroism."[33]

Revealing the reality of wartime society in a critical tone was the primary function of works published in the journal. The focus for selecting works, according to chief editor Hu Feng, was whether or not the author could "seize the essential elements and avoid wordy, flat or exaggerated description, possess a critical spirit and bravely expose dark or dirty phenomenon, and discard conceptual, abstract discourse in their use of language."[34] In December 1937, Hu pointed out two weaknesses in the wartime reportage genre—flat description and excessive sentimentalism. Hu argued that to remedy these shortcomings, writers should express their emotions through concrete details selected from real life experiences and depict national heroes as complicated human beings with both strengths and weaknesses.[35] The independent nature of *Qiyue* and its spirit of social criticism allowed its contributors, including Qiu Dongping, to break the tough and well-disciplined soldier ideal and to remove the glorious aura of bravery, sacrifice, and military discipline advocated in the Nationalist discourse.

The editor's directives set the tone for the style of Qiu Dongping's battlefield reportages—"personal" and "critical." Qiu's ability to provide a personal and critical description of soldiers' combat stories also came from his direct army experiences. Born in Haifeng, Guangdong, Qiu Dongping herded cattle in his village as a teenager and studied at a middle school at sixteen. He joined the CCP in 1927, participated in a peasant uprising in Haifeng in 1928, and escaped to Hong Kong after the failure of the uprising. In Hong Kong, he supported himself by fishing, doing various odd jobs, and managing a small business in his spare time. In the meantime he started to practice writing.

After Japan invaded the northeastern parts of China in 1931, Qiu went to Nanchang, Jiangxi, to meet his elder brother Qiu Guozhen (1894–1979), who served as the chief of staff (*canmouzhang*) in a brigade of the Nineteenth Route Army of the Nationalist military forces. Qiu joined the Nineteenth Route Army and fought against the Japanese in the 1932 Battle of Shanghai. After the Shanghai Ceasefire Agreement was signed in May, the leadership of the Nineteenth Route Army revolted against the Nationalist government in Fujian Province. In response, Qiu left the army and went to Hong Kong to launch propaganda campaigns for national salvation. He later participated in the Second Battle of Shanghai in August 1937 and went on to join the CCP's New Fourth Army, led by Ye Ting (1896–1949) in the spring of 1938. He assisted the CCP leader Liu Shaoqi (1898–1969) in operating a branch of Lu Xun College of Arts and Literatures (*Lu Xun yishu wenxue yuan*) in central China. In 1941, however, he was killed by the Japanese during a skirmish in Yancheng, Jiangsu, during a botched attempt to help college students break out of a Japanese encirclement.

With the personal experience he gained from serving in the army, Qiu described in great detail the soldiers' everyday lives and fighting experiences, as well as their emotions and mind-sets. He adopted a critical approach to soldiers' combat experiences and conveyed his personal opinions "either explicitly through authorial comments or implicitly through the selection of details."[36] This approach allowed his battlefield reportages to de-idealize war heroics through relentless exposure to the combat trauma of Nationalist army soldiers.

Qiu's first major contribution to *Qiyue* was the reportage "The Seventh Company" (*Diqilian*), published in January 1938 in the journal's sixth issue. Drawn from the real story of his younger brother, who was injured at war, "The Seventh Company" narrates cadet-turned-commander Qiu Jun's experience of modern warfare. Qiu Jun is a graduate of the Guangzhou branch campus of the Central Army Officer Academy. During the Second Battle of Shanghai, he is sent to the front and is appointed commander of his company. The reportage describes in detail the psychological transformation that Qiu Jun experiences as he matures from a military academy graduate to an army officer. Qiu Dongping does not confine the focus of his writing to describing the events or behaviors of his characters. Instead, he delves into the mind and subconscious of Qiu Jun.

"The Seventh Company" starts with a description of Qiu Jun's fear, self-doubt, and total incomprehension of the battlefield under heavy barrage. In this reportage, even a professional army officer who has received formal military training—like Qiu Jun—comes to realize the extent of his own faintheartedness during battle. Before the battle starts, Qiu doubts the strength of his willpower, worrying that he might be the most cowardly among his military academy classmates. To strengthen his fighting spirit, he keeps himself away from elements that, in his view, might crumble his strength as a soldier:

> After we set out from Kunshan, I began a solemn and strange journey. Along the banks of a little creek near Qianmentang, a young, beautiful woman wearing a green dress appeared before our ranks. To all the men I said: "Halt, let's take a rest here!"
>
> Chen Weiying, a platoon leader, quietly asked me: "Why do you want to stop? Let's catch up with her; what could be wrong with walking next to her?"
>
> "This is my own philosophy," I said. "Now every time when I run across a pretty woman I steer clear because she will stir up a lot of unnecessary, harmful ideas in me. . . ."
>
> Our special operations officer (*tewu zhang*) brought a phono-

graph from Taicang, and I made him hand it over to me; I took all of his records and smashed them because I am afraid to listen to music, too.

I constructed my own path with extreme care, as if I were cutting away brambles and paving with stones—because I want to make myself into a proper soldier, so that I can stand firm on this momentous battle line. On every side I was protecting myself from the poisons of emotion.[37]

As an army officer who has received formal military training, Qiu Jun has a strong sense of professional responsibility and honor. As he narrates, he plans his career path carefully and tries to appear as a proper soldier. Because of Qiu Jun's professional pursuits, he faces head-on his susceptibility to the distraction of personal desires and consciously tries to avoid them.

For a professional army officer like Qiu Jun, the enemy is not just the Japanese but also the distraction of personal emotions, which are condemned by Qiu as poison that would weaken the fighting will of a combatant. These emotions that Qiu tries to restrain include not only attraction to women and music but also fear. The author vividly depicts Qiu Jun's anxiety and fear when his troops stationed at the fire line are faced with overwhelming gunfire by the enemy. The intimidating description of combat brutality removes the glorious aura surrounding the cult of physical sacrifice advocated by Chiang Kai-shek's speeches to Whampoa cadets. Qiu Dongping writes:

The enemy artillery fire was amazingly accurate and their cannon shots followed and chased our routed soldiers closely and relentlessly like a group of spirited, running ghosts. Having thrown away their weapons, the [Chinese] soldiers, covered with blood and mud, fled in the dense black smoke like mad wolves. The enemy gunfire was fierce and it appeared all the more fierce when it created fear on the battlefield and forced our soldiers on the front line to retreat helter-skelter and pitifully, creating a frightening picture we had never seen. It not only confused our morale, it almost completely snatched away our morale. I realized that this frightening scene alone could dissipate our fighting spirit before the enemy gunfire destroyed us.[38]

The brutal and overwhelming enemy fire appears several times in this reportage. The narrator, Qiu Jun, comments that the battle scenes become fragmented and horrible impressions for the soldiers, whose minds are disoriented in combat. He describes the uniquely deafening sound from the enemy gunfire as "a mysterious and horrible world," which has made him "immersed in depression."[39] Qiu Jun's vivid depiction of battle scenes reveals his frustration and helplessness: "The enemy's accurate mortars played a cruel joke on our Chinese army's battle formation. The curved line created by the mortars' impact ruts was a mirror image of the curved line formed by our skirmishers. The dense artillery fire made the ground around us shake differently; it no longer vibrated like a spring but seemed to be dissolving the earth, like eruptions of massive waves in a bottomless sea."[40] The battle is described as a world totally strange to a soldier who received formal military training. Qiu Jun comments, "It seemed that I did not remember the gunfire; neither did I know it."[41] The overwhelming fire causes severe fear among soldiers: "We crouched in our trenches, grinding our teeth, enduring the irresistible weight of the artillery fire. In the beginning, we were calculating the rest of our lives in terms of months and weeks. Gradually it became days, hours, seconds, and now it was thousandths of a second."[42] Qiu Dongping's detailed description of the terrifying battle and the soldiers' fear contradicts the Nationalist propaganda for the soldier figure. This reportage does not celebrate the glory of death during battle as forged in Chiang Kai-shek's speeches to Whampoa cadets. The injuries and death caused by the brutality in battle are stripped of their romantic and glorious aura.

For Qiu Dongping as a writer with personal army experience, a proper soldier possesses enough personal responsibility to conquer his fears. The reportage shows that Qiu feels it pointless to differentiate between what is bravery and what is cowardice; all he remembers and ponders in his mind is his combat task.[43] Qiu Jun's sense of professional duty as a soldier drives him to unconditionally follow his commander's order to "stand or fall with your formation!" The reportage reads:

> Our regimental commander called me on the phone, asking me bluntly: "Can you hold out?"
>
> "Yes, commander, I can," I answered.
>
> "I hope it is perfectly clear to you that this is your chance to do something important and make a name for yourself; you must

be deeply conscious of our righteous cause and be determined to stand or fall with your formation!"

I felt as if my commander were speaking directly to my soul; his words (according to the Chinese manner of communication between humans and spirits) should have been written on paper and burned—and I was moved from the depths of my soul by his words, moved to the point of tears.[44]

Qiu Jun considers standing or falling with the formation to be the inherent nature of his profession. As the Chinese-literature scholar Charles Laughlin comments, "Qiu Jun overcomes his spiritual struggle with desire and emotion by recasting the stereotyped relationship with the battle formation as a projection of his very conscience and sense of responsibility into those very trenches."[45] Laughlin's comments support the view that it is Qiu Jun's conscience and sense of responsibility as a professional soldier that helps him overcome the distraction caused by fear.

Although Qiu Jun's sense of professional duty as a soldier drives him to unconditionally follow his commander's order, he cannot help himself from critically thinking about the order. No matter how moved Qiu Jun feels about the order to stand or fall with his formation, he thinks that this order does not make much sense: "'Stand or fall with your formation!' I was calm. I was constantly protecting myself from being cheated by these words. I felt the sentence was entirely wrong: Chinese generals and officers loved using it, and I was well aware of the sentence's sacred significance, but I was still afraid that I would be swindled by it somehow. One's 'standing' or 'falling' was really not an issue here; the guarding of a formation was another matter, which was really more important than one's standing or falling."[46] As a professional soldier, Qiu Jun understands that the defense of the battlefield is beyond the soldier's control. Whether or not the soldier successfully completes his military order is less important than the strength of his resolve as translated into arduous actions in defense of the nation. Qiu Jun, as an army officer who has received professional military training, meets the Nationalist ideal of a disciplined soldier in that he unconditionally follows his commander's order. However, unconditionally abiding by military orders is only part of his sense of professional responsibility, as he also ponders the order from a critical perspective. A model soldier for Qiu Dongping is not simply a fighting machine totally controlled by the army; he should maintain a degree of intellectual independence. In this sense,

Qiu Dongping questions the Nationalist ideal of a disciplined soldier. By depicting the hero of his fiction as an army officer who has a questioning mind and maintains intellectual autonomy, Qiu Dongping also asserts his intellectual independence as a writer and social critic.

Qiu Dongping's questioning of the Nationalist ideal of soldier heroics is also manifested in his critical understanding of bravery and soldiers' bonding, the virtues advocated in Chiang's speeches and Whampoa's motto. As a professional army officer, Qiu Jun appreciates the significance of comradeship in combat. As Laughlin comments, "the arduous and tragic process of building the trenches is also a process through which Qiu Jun forges a caring, human relationship with his men."[47] At night he walks on the battleground alone, and when he sees his soldiers sleeping, Qiu finds consolation from the thought that he is able to build friendship with them.[48]

Although Qiu values emotional bonding with his soldiers, he does not let it prevent him from making rational decisions. When a platoon leader under Qiu's command is about to leap from a foxhole to support his comrades, Qiu Jun stops him. In his eyes, "unnecessary exposure to enemy fire in the wake of their hasty attacks" is irrational behavior.[49] He views such an action as misguided bravery and "impetuous heroism," only capable of bringing meaningless casualties while presenting little to no benefits to the battle.[50] But the platoon leader does not follow Qiu's order; his action "inexorably exposed the configuration of the company's battle formation to the enemy, whose cannon fire increasingly mirrored the actual shape of the formation."[51] When this platoon leader realizes he might be shot because he has violated Qiu Jun's order, he deserts in a panic.

Qiu Dongping also subverts the Nationalist ideal of soldier heroics by revealing that civilians did not necessarily treat the soldier as a model citizen to emulate. Instead, they thought of the soldier as an information source to satisfy their curiosity of war, as soldiers and civilians had different understandings and experiences of war. When Qiu Jun leaves for the front, his nephew gives him a leather satchel and asks him to put a piece of an enemy soldier's skull, an enemy army flag, and parts of an enemy machine gun inside the satchel. This request shows the romantic sentiment toward war and soldiers among civilians. For Qiu Jun's nephew, a war hero is manifested in his abilities to kill the body of the enemy, to destroy all the enemies and seize their flag, and to use modern weaponry. However, Qiu Jun does not respond with any heroic utterance to his nephew's request; instead, he feels his nephew's thoughts are absurd and ridiculous. War, which is unpredictable even to professional soldiers like him, is certainly beyond the comprehension of civilians like his nephew. Qiu Jun is not concerned about being a model citizen for the civilians; he only aims to prove to himself that he is a proper soldier who fights bravely, unconditionally follows orders, and is also capable of maintaining his own intellectualism in order to contemplate given orders.

The soldier's unique war experience in Qiu Dongping's depiction centers on the paradoxical meanings of war and violence. On the one hand, it is violence that gives the soldier a sense of strength and victory. In the 1935 reportage "Honghuadi zhi shouyu" (The defense of Honghuadi), Qiu writes: "The enemy could not realize his upcoming fate even just a thousandth of a second before he died, and his fate was exactly controlled in the soldier's hands. The soldier thus developed a deep comprehension of the true essence of what was strength and what was victory. And this was the fortune that only soldiers could appreciate."[52] The soldiers draw power from killing—a sense inaccessible to civilians. On the other hand, it is also war violence that makes the soldier aware of his faint-heartedness and vulnerability. Qiu Jun in "The Seventh Company" fails the heroic ideal of committing suicide in the event of a total defeat on the battleground. Nor does he fulfill his promise to put a piece of an enemy soldier's skull, an enemy army flag, and parts of an enemy machine gun into the leather satchel as his nephew asked. When he returns from the battlefield with nothing but failed promises, he gains little more than the realization of his own frailty. War provides a source from which the soldiers draw their power, but it also reveals their weakness and causes scars that will hardly heal.

The paradoxical meanings of war to soldiers and civilians are highlighted in Qiu's 1932 reportage "Tongxunyuan" (Correspondent). The story begins with a villager, Lin Ji, who is appointed as a correspondent because he killed a fat rent collector and thus is considered the bravest person in the village. Yet the reason of the killing is not explained, and it appears unnecessary for the killer and the bystanders to justify killing. What matters is the contrast of the spectators' attitudes toward Lin before and after killing. Before the killing, Lin is mocked by villagers who question his courage to use a gun to kill.[53] After the killing, however, Lin is praised by the villagers.[54] The violence of killing is considered by civilians to be a myth which they are eager to observe, imagine, and discuss. Lin's neighbors tirelessly talk about their imaginations and speculations of war. They are eager to inquire about "some secret matters" from Lin and take delight in "spreading rumor and hearsay about war."[55] In this manner, they conceive of smart and resourceful combat strategies as if they are fighting in a battle. War and violence do not mean trauma, scar, or fear to the civilians who are uninfluenced by it; instead, they are just legendary stories used for entertainment.

Lin Ji's response to the neighbors' fantasies of war is "only smiles with few replies."[56] For the soldier Lin, war does not have any romantic or glorious aura. When Lin Ji escorts a young soldier to perform a task, the young soldier is captured and brutally killed by the enemy because he lacks enough battle experience to overcome his attackers. Over time, Lin Ji feels it almost impossible to forget about the young soldier's tragedy. Although Lin's commander does not blame him and instead offers sincere comforts to him, he is still haunted by the scream of the young soldier and is thus trapped in a pain that cannot be alleviated. Lin blames himself for his inability to save the young soldier, and eventually suffers mental derangement and commits suicide. Lin's story questions the Nationalist discourse that transformed the gloomy and dark sides of death into a glorious and solemn beauty. The sacrifice of Lin Ji's fellow soldier is not considered by Lin as an honor of eternal life; instead, it is a tragedy.

As a writer with personal combat experiences, Qiu Dongping reveals not only the soldiers' trauma in combat but also the absurdity and irony of the war. In the reportage "The Seventh Company," when the twenty-five soldiers who survive are madly and dizzily seeking their enemy in the midst of gunfire, soldiers from their allied army misidentify them as the enemy and shoot at them.[57] The war is ironic in Qiu Dongping's reportage also because military discipline, which was highly celebrated in the Nationalist discourse for making model soldiers, is stripped of its aura. One example is the 1938 reportage "Zhongxiao fuguan" (The lieutenant colonel), which details the combat experience of a competent middle-level army officer who respects his commander but is shot dead because he questions the commander's unreasonable order.

The lieutenant colonel highly respects and trusts his army commander, treating the commander as his idol. He believes that military generals can be viewed as sacred flags that symbolize national pride and honor. In battle, he is a competent officer who not only fights bravely but also has deep concerns about national resistance. For instance, he feels grief when he sees the death of common people due to their ignorance of war. He also detests the irascible attitude of the soldiers when they are educating the common masses on self-protection when under enemy bombardment. The lieutenant colonel comments that if the common people had been mobilized into the national salvation campaign earlier, they would have the intellect to understand the war and violence.[58]

His competence as an army officer is also manifested in his critical thinking regarding military orders. When the army commander orders him to retreat, claiming that retreat is the most strategic option, he instead holds the position and keeps fighting to support the allied army that is still engaged in combat. The first time he questions his army commander is over the question of retreat: "Does the military strategy teach a soldier to abandon the territory of the country?"[59] Under the order of the army commander who is "sacred and inviolable," the lieutenant colonel is shot dead.[60] Qiu Dongping's description of the face of this dead soldier who dares question the commander's retreat order is full of respect: "His face, which is as strong as a horse, is like a perpetual statue of an old sage covered with steadfast and calm smiles and wrinkles. This expression that can only be seen in the faces of combat soldiers connotes his question of the military discipline in the army and battlefield."

Qiu Dongping expresses his respect for the lieutenant colonel, for fearlessly fighting against the enemy and bravely expressing his critical opinion of the commander. When allied forces finally defeat the enemy the next day, the army commander—who has just killed the lieutenant colonel—makes the false, empty claim in his report that "I will not lose any inch of the land with the determination to sacrifice my life."[61] This report reveals the ironic aspect of military discipline in the Nationalist army. A soldier who has the bravery and resolution to fight to the death is not killed by the enemy but by his own commander, whom he highly respects. The destiny of the soldier does not solely depend on the consequence of the battle; it also depends on whether the soldier absolutely obeys his commander's order, even though the order is to give up and retreat.

The trauma of soldiers and the absurdity of war are also revealed in the 1938 reportage "Yige lianzhangde zhandou zaoyu" (A company commander's combat experience). It was acclaimed by Qiu's fellow writers as the best specimen of wartime reportage.[62] In this reportage, Qiu Dongping describes the mental pain and emotional scars of a company commander while criticizing the rigidity of military discipline in the Nationalist army. The commander of a company, Lin Qingshi, meets a friend from his military academy, Gao Feng, who retreated from the front. Gao's description of his combat life is filled with misery and reveals a gap between his professional goal and the war's cruel reality:

> I think that all the soldiers are miserable. When a soldier graduates from the military academy, he wears a sword and military uniform, looking as brave and strong as other soldiers. When walking in the street, he often attracts the admiration from others. . . . When he joins the front, death and injury is not his concern at all; the most painful thing to him is the inability to complete the combat task. I have great ambition to be an excellent soldier, which even sounds arrogant to others. Because of my ambition I always respect myself and feel proud of myself in front of others. . . .
>
> Three months ago, I served as a second lieutenant at an army unit in Guangdong. My wife and my friends wrote letters to me congratulating on my promotion. However, I do not think it is my honor. I feel that I am marching in thick fog. My trace is so secret that nobody knows where I come from and leave for next. . . . Not very long ago, our troops marched to the front and I served as a platoon leader. I knew that I might grow into an excellent soldier in combat. . . . On the night of November 18, we were attacked by a group of powerful enemies. Thirty-five soldiers died, and I was the only one that survived. This fact has extremely shocked me. I cannot figure out what the combat is now. I feel that the combat is like a robber or a thug. Whenever you relax just a little bit, it confronts you immediately. What has made me feel most painful is that ever since the combat started, we were confined to the passive status of being attacked. Our guns were held in our hands, but we just could not find our opponents. . . . The combat is serious. I think that I have recognized its solemn yet cruel face. This face makes me scared and I really do not have the bravery to confront it.[63]

Gao Feng's narrative demonstrates how the ambition of a military academy cadet to be a great soldier inevitably will be compromised by the brutality of war. No matter how well a soldier is trained at the military academy and how firm his fighting resolve is, the brutality of war easily disorients him. His promotion, which was admired by civilians, does not heal his mental loneliness and emotional scars.

In this work, the soldier's war trauma is not only caused by powerful enemy fire but also from the sheer rigidity of military discipline. Lin Qing-shi, touched by his soldiers' determination to fight, decides to break from the order from his commander and launch his troops into battle against a group of well-equipped Japanese enemies. When Lin makes this decision, he knows that he will be executed by his commander even if he wins the battle. After victory, Lin surrenders himself to his commander to take full responsibility for disobeying the order. Qiu Dongping makes it clear that Lin's surrender is motivated by his sense of professional duty. He writes: "Lin could have escaped from the punishment, but he did not defend him-self at all. Instead, he chose to accept the execution by the military law just to fulfill his dignity as a soldier."[64] Qiu Dongping laments: "Unfortu-nately he is not defeated by the fierce fire of the Japanese army but killed by his fellow soldier."[65] Qiu's comment criticizes the rigidity of military disci-pline in the army. Unconditional subordination to the military hierarchi-cal structure may not result in victory in battle. Critical thinking allows the competent army officer Lin Qingshi to make a decision beneficial to vic-tory. Nonetheless, Lin's choice, regardless of the military outcome, leads to his execution. This reportage does not stop with Lin's execution; instead, it ends with Qiu Dongping's comment, "New Chinese soldiers are rising."[66] These new soldiers Qiu is referring to are those like Lin Qingshi and the lieutenant colonel detailed above, who not only bravely and competently fight against the enemy but also possess the ability to critically analyze mil-itary situations and maintain intellectual independence.

Qiu Dongping's battlefield reportages published in *Qiyue* provide per-sonal and critical descriptions of soldiers' trauma and the absurdity of war. Several factors led to Qiu's personalized and critical writing strategy. *Qiyue* was a literary journal that maintained its independence instead of serv-ing as a propaganda mouthpiece. The journal's editor, Hu Feng, selected reportages based on how well they critically and realistically revealed the dark sides of wartime Chinese society. Qiu Dongping's personal army and combat experiences allowed him to develop a critical and deeper under-standing of the war and the soldiers' combat experiences.

With a personal and critical tone, Qiu's battlefield reportages reveal traumatic experiences of soldiers during the war and attack the absurdity and irony of the war. Although soldiers are depicted as model citizens and the epitome in the GMD's political rhetoric, that image is heavily contra-dicted by the soldiers in Qiu's reportage writing. In his works, the soldiers were the ones who greatly suffered from and were subsequently traumatized by the horrors of war; it was difficult for civilians to understand the soldiers' plight. By revealing the reality of war's brutality and soldiers' trauma, Qiu Dongping subverted the Nationalist discourse of the soldier figure by removing the romantic and glorious aura that surrounded the sacrifice, bravery, and military discipline. In doing so, Qiu asserted literary intellectuals' role as social critics while advocating critical thinking as a great virtue for Chinese soldiers. Like Xiao Jun, Qiu Dongping expressed deep social consciousness by exploring the dark sides of wartime Chinese society and searching for the hope of national salvation. In Xiao's fiction, the hope resided in grassroots soldiers who bravely fought for peace (or revenge) and to better their family life. In Qiu's battlefield reportage, hope mixed with a strong sense of professional responsibility that motivated the soldiers to fight fearlessly and to maintain critical thinking and intellectual independence.

The Female Soldier Writer Xie Bingying

The soldiers in the writings of Xiao Jun and Qiu Dongping are all males, either guerrilla soldiers or professional army officers. This section examines the autobiographies of female Whampoa cadet Xie Bingying (1906–2000) in an attempt to show that soldier imagery in literary works could also be gendered. Soon after the Whampoa Academy was founded in 1924, a debate on whether Whampoa should accept women as cadets unfolded. A cadet named Li Zhilong (1899–1928) published an article about this issue on August 17, 1925, in the *Zhongguo junren* (Chinese soldiers), a Whampoa journal published by the Association of Chinese Young Soldiers (*Zhongguo qingnian junren lianhehui*), led by Chinese Communists at Whampoa.[67] This article, titled "Lujun junguan xuexiao zhaoshou nüsheng wenti" (The issue of recruiting woman cadets into the army officer academy), supported the view that Whampoa should allow women to join the academy and become soldiers. Li pointed out that women were suitable for many roles in the army, such as staff officer (*canmouguan*), party representative, political worker, intelligence worker, logistics worker, and communications worker (*tongxinyuan*). He opposed recruiting all women into the army since he believed that women were physically weaker than men, but he advocated that as long as the women who wanted to join Whampoa had strong bodies, they should be allowed to join.

Li also made public a letter written by a woman named Jin Huishu[68] to the GMD party representative of Whampoa, Liao Zhongkai, in which Jin requested that Whampoa recruit female cadets. Jin argued that it was a sign of inequality between men and women if women were barred from army service. In response, Liao expressed his understanding and support for Jin's request. He also spoke against an article published in *Guangzhou ribao* (Guangzhou daily) on July 13, 1925, that criticized recruiting female cadets. The article argued that women's presence in the army would cause sexual disorder and weaken military morale. Liao suggested that as long as women were physically strong, they should be allowed to enter Whampoa.[69]

Liao allowed Whampoa to recruit a small group of women cadets on the condition that he expected the women to be strong and perform assistant tasks, such as political work support. This decision was made as a compromise to promote equal rights for women and was intended to build the prestige of the newly founded Whampoa. After the Northern Expedition began in 1926, the Whampoa Military Academy opened a branch campus in Wuhan—the Wuhan Central Military and Political Academy (*Wuhan zhongyang junshi zhengzhi xuexiao*). This branch campus recruited female cadets for the first time in the history of modern China.[70] Two hundred and thirty-one women were accepted and constituted part of the sixth class of the Whampoa Military Academy.[71] The women received training from February to July 1927 and were offered full scholarships that covered all tuition costs and food and accommodation expenditures as well as giving the women a monthly stipend.[72]

This section focuses on the writings by one of the Whampoa female cadets, Xie Bingying, and examines the literary construction of the image of the soldier figure from a female perspective. Xie was one of the first female soldiers in modern Chinese history and also is well known for her literary production. Her autobiographies *Congjun riji* (Army diaries, 1928), *Xin congjun riji* (New army diaries, 1938), and *Yige nübingde zizhuan* (The autobiography of a female soldier, 1936) received overwhelming praise in China and abroad when they appeared. They have been reprinted multiple times and have been translated into several languages.[73] Xie's works serve as a useful window to examine the motivation and mentality of female soldiers in modern Chinese history.

Xie was a rebellious child who had a strong sense of bravery and determination and eventually grew into a soldier and writer. She portrayed the female soldier as a woman fighting to free herself from the constraints of traditional Chinese society, especially the gender roles in marriage. Born into a traditional family, she was expected to be an obedient daughter and, later on, daughter-in-law. Her education was largely restricted to learning how to spin cotton and embroider, and her reading was limited to such books as *Jiaonü yigui* (Teach your daughter traditional rules).[74] Xie's rebelliousness was often criticized by her mother as exemplifying a lack of daughterly obedience. From childhood, Xie was eager to control her own fate and fought hard to do so. She tried her best to resist the conventional gender-based training she received from her mother. Despite her mother's scolding, Xie ventured outside to play with the boys instead of going into the fields with older, engaged girls to pick tea leaves during the day and weaving in the evenings to prepare her trousseau. She "fought fiercely against having her feet bound by begging, coaxing, kicking, snatching, crying and roaring."[75] Xie recounted her efforts to secure a formal education at a *sishu* (private school) at the age of ten by threatening to commit suicide.

Xie brought her rebellious spirit to her school and later on to society. At twelve, she led a student uprising at a girls' school because the instructor told her not to talk and play with boys. Later she was expelled from a church-based high school because she organized a student council with several other students and initiated a student parade to commemorate the 1925 May Thirtieth Movement, a major labor and anti-imperialist demonstration. She then joined the Wuhan branch campus of the Whampoa Military Academy in 1926 at the age of twenty. Her rebellious personality and her heroic and chivalrous spirit were some of the main reasons she joined the army. She wrote in her autobiography: "Among the traditional novels, *Shuihu zhuan* (All men are brothers) is my favorite one. Although *Honglou meng* (Dream of the red chamber) is also a masterpiece, it could not arouse my interest as *Shuihu*. I dislike the crying of Lin Daiyu. I also don't like the foolish-looking (*shatou shanao*) of Jia Baoyu, who spent his whole day hanging around girls. I truly admire every hero in *Shuihu*. Their brave and chivalrous spirits exerted great influence on my later participation in the army."[76]

Xie's narrative of her chivalrous spirit collaborated with the heroic soldier ideal advocated in Chiang Kai-shek's civic education speeches at the Whampoa Military Academy. She wrote that the female cadets at Whampoa washed the rouge and powder off their faces and cut their hair short.[77] Xie believed that to be female soldiers meant to cleanly rid themselves of all feminine qualities and to emulate the appearance and behaviors of male soldiers.[78]

However, Xie did not echo Chiang Kai-shek's celebration of the heroic soldier ideal with the intention to reform the army or to strengthen Chiang's authority. Instead, Xie's heroic soldier ideal aimed to justify her challenge of traditional gender roles and her struggle for individual independence. Xie mentions in her autobiography that she joined the army because of the pursuit of her writing career and her wish to control when and whether she married. She believed that a writer could not produce powerful and lifelike works without having some unique life experiences, and she saw joining the army as a great opportunity to build her physique and acquire substantial materials for her writing.[79] Moreover, joining the army provided a means for her to escape the marriage arranged by her mother. She wrote in her diary that the top reason that female cadets joined the army was to "escape from the oppression of the feudal family."[80]

For Xie, the heroics of a woman soldier meant not only fighting against traditional practices and values that confined women within the patriarchal family but also meeting any life difficulties women faced in general. After the collapse of the First United Front between Chinese Nationalists and Communists in 1927, the class of female cadets was disbanded by Whampoa, forcing Xie to return home and fight against the marriage her family arranged for her. She experienced the failure of her first marriage and became a single mother. She was also thrown into a prison because she refused to attend the ceremony welcoming Puyi (1906–1967), the emperor of the puppet state of Manchukuo sponsored by the Japanese.

Xie echoed the GMD's rhetoric of the soldier not only in celebrating the heroic soldier ideal but also in claiming assistant roles in the army for women soldiers. Xie's autobiographies show that she did not challenge Whampoa's expectations for female cadets to perform noncombat assistant tasks. In her autobiographies, she often propagated meanings of the Northern Expedition to social masses who had little knowledge.[81] Her main task during the Northern Expedition was to provide treatment to wounded soldiers, but she made political work her priority.[82] As revealed in chapter 3, during the Second Sino-Japanese War, Xie left her home and traveled to Changsha to organize the Women's War Service Corps, which went to the front line to provide treatment for the wounded soldiers and make war reports.

Although Xie collaborated with the GMD state in performing noncombat tasks, her goal of performing political work was not just to propagandize nationalist ideology and foster the patriotic spirit among the social masses. In a letter she wrote to her brother, she listed her threefold agenda:

to treat and encourage the wounded soldiers, to perform political propaganda by mobilizing the peasants and workers to participate in the revolution, and to educate the women in villages and factories to fight for their liberation. Xie considered mobilizing women to fight for liberation and against patriarchal confinement to be an important goal in her political work. She argued that women's active participation in the nationalist revolution was the manifestation of China's modernity by citing examples of women's service in the revolutions in France and Russia.[83] She was excited to see more village women holding their own local meetings and "fighting against the oppression from the feudal and bourgeoisie forces."[84] As Xie's friend Yi Ming[85] commented, Xie Bingying considered participating in the war service to be the only way for women to truly achieve emancipation.[86] Although Xie did not challenge the GMD's ideal for a female soldier, her goals in performing political work differed from the GMD's expectations for female soldiers.

This chapter has examined the literary images of three types of soldiers: soldiers in a rural guerrilla unit, soldiers in the Nationalist army, and a female soldier who studied at Whampoa. It has also shown how Xiao Jun, Qiu Dongping, and Xie Bingying created gendered images of soldiers in the literary works that represented the perspectives of male and female writers. These three writers all had army experiences, either as a military academy cadet or as a combatant. Therefore, it is safe to argue that they had deeper and more personal understandings of the war and of the mentality of soldiers than civilians and other writers. These literary intellectuals who had direct army experiences either subverted or, conversely, supported the Nationalist rhetoric of the heroic soldier figure in order to assert their social consciousness and serve their goals.

Xiao Jun's fictional work *Bayuede xiangcun* and Qiu Dongping's battlefield reportage subverted the Nationalist rhetoric by depicting soldier heroes as ordinary human beings who had personal motivations to fight and who suffered from the war. The guerrilla soldiers in *Bayuede xiangcun* were not emulating the Nationalist ideal of politicized, disciplined, and morally cultivated citizen-soldiers. It was misery caused by the social crisis and the Japanese invasion that forced them to participate in the war. The soldiers, armed with weapons, only enjoyed an illusory sense of honor and strength. In actual combat they severely suffered from the war, experiencing frustration and helplessness. The brutality and absurdity of war and the trauma soldiers endured were hardly understandable to the civilians. The writers' combat stories removed the glorious aura surrounding death, bravery, and military discipline that Chiang fashioned in his speeches.

Several factors could explain why the literary works by Xiao Jun and Qiu Dongping subverted the heroic soldier ideal promoted by the GMD. These include the tough personalities of Xiao and Qiu; their direct experiences with the military academy, the army, and combat; the personal and critical style of their writing; and their concerns for the misery of the peasants, common soldiers, and lower-level army officers.

When depicting the soldier figure in their works, Xiao and Qiu set their aims higher than patriotic agitation or ideological propaganda. In subverting the Nationalist rhetoric of the heroic soldier figure, Xiao Jun and Qiu Dongping both extended their thinking of China's national crisis into creating the soldier figure. In doing so, they asserted the role of literary intellectuals as social critics. They treated the army as a microcosm of Chinese society, exploring the dark phenomena and the roots of the common people's suffering. They attacked the cruelty of the enemy, the brutality of the war, the corruption within the army, and the rigidity of the military hierarchy. In attacking the brutality of war and the soldiers' suffering, Xiao Jun and Qiu Dongping were also thinking about how to achieve China's national salvation. Xiao revealed the potential martial strength of rural peasants, who had strong fighting spirits. Qiu advocated that a soldier's sense of professional responsibility included not only fighting fearlessly but also practicing critical thinking and intellectual independence.

The literary images of the soldier figure were highly gendered. In *Bayuede xiangcun,* male soldiers fought the war to protect their families while the female soldier was forced to join the unit after experiencing severe misery. The suffering of soldiers also was gendered, as sexual oppression constituted a significant part of women's war trauma. Although Xiao and Qiu subverted the heroic soldier ideal, the female soldier Xie Bingying echoed this ideal by highlighting her chivalrous spirit. Xie also collaborated with the GMD in performing political work rather than combat tasks. However, her collaboration allowed her to achieve her personal agendas, which were to break the confinement of traditional gender roles and to argue for women's political participation. As her autobiographies show, her military service at the Whampoa Military Academy was a continuation of her fight for an independent lifestyle and her commitment to social service.

The symbol of the Chinese YMCA Emergency Service Programs to Soldiers is a red triangle with a blue strip in the middle. Their slogan is "To serve as the bridge connecting the military and the civilian, and to be friends with soldiers."

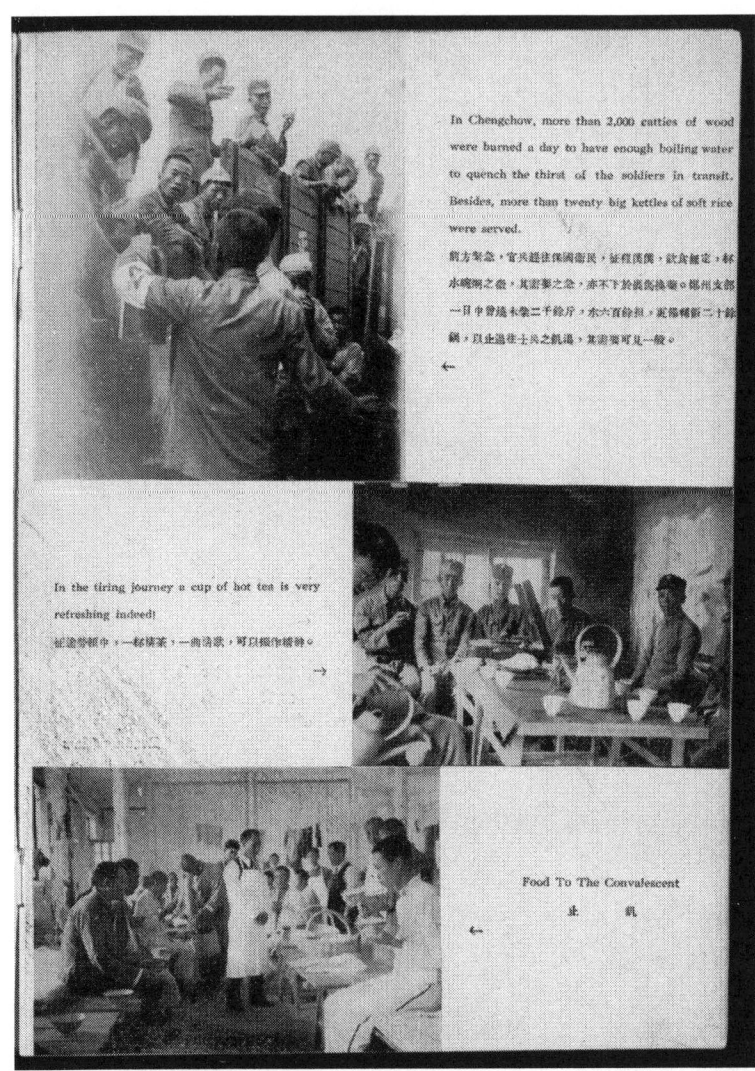

In Chengchow, more than 2,000 catties of wood were burned a day to have enough boiling water to quench the thirst of the soldiers in transit. Besides, more than twenty big kettles of soft rice were served.

前方緊急，官兵趕往保國衛民，征程浩蕩，飲食無定，杯水嘯渴之急，其需孰之急，亦不下於槍砲換藥。鄭州支部一日中嘗燃木柴二千餘斤，赤六百餘担，更備稀飯二十餘鍋，以此过往士兵之飢渴，其需要可見一斑。

In the tiring journey a cup of hot tea is very refreshing indeed!

征途勞頓中，一杯清茶，一曲清歌，可以振作精神。

Food To The Convalescent

止　飢

Chinese YMCA workers providing hot tea and food for soldiers. The text on the right top reads, "In Zhengzhou, more than 2,000 catties of wood were burned a day to have enough boiling water to quench the thirst of the soldiers in transit. Besides, more than twenty big kettles of soft rice were served." The text in the middle left reads: "In the tiring journey a cup of hot tea is very refreshing indeed!"

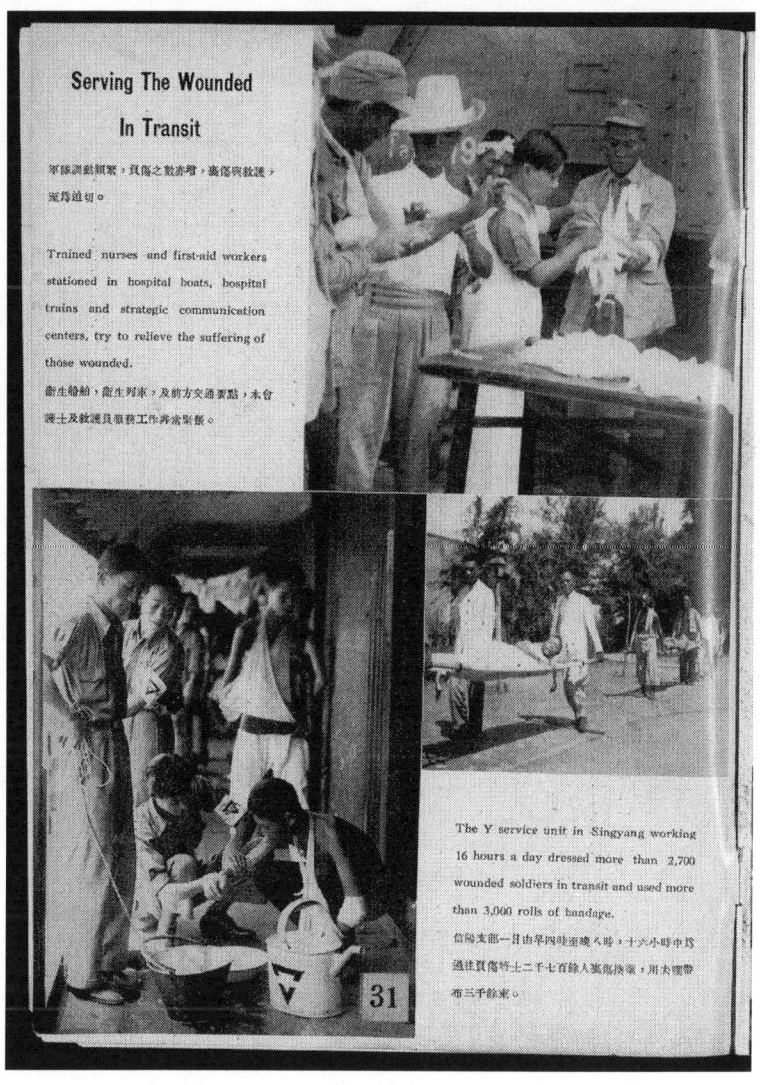

The Chinese YMCA serving wounded soldiers in transit. The text on the top reads: "Trained nurses and first-aid workers[,] stationed in hospital boats, hospital trains, and strategic communication centers, try to relieve the suffering of those wounded." The text on the bottom reads: "The Y service unit in Singyang [Xinyang in Henan Province] working sixteen hours a day dressed more than 2,700 wounded soldiers in transit and used more than 3,000 rolls of bandage."

我們好比上火線！　　沒有後退祇向前！

Teaching Singing to keep Up Morale

教 唱 歌　　振 士 氣

起 來 ！　　起 來 ！　　起 來 ！

Chinese YMCA workers teaching soldiers songs to maintain morale.
The lyric reads: "We are marching to the front line! We will never
retreat!"

Writing Letters for Wounded Soldiers

替傷兵寫家信

"Please tell my mother...."

「請告訴我的母親……」

Distributing Comfort Materials

分贈慰勞品

Lectures

演講

Chinese YMCA workers helping wounded soldiers write family letters, distributing comfort materials, and delivering educational lectures.

Serving the Wounded in Various Ways

Soldiers And Civilians Are

Serving Together

軍　民　聯　歡

Chinese YMCA workers organizing a party for wounded soldiers where various genres of art were performed.

Chinese YMCA workers organizing ping-pong, chess, basketball, and other recreational and sports activities for soldiers.

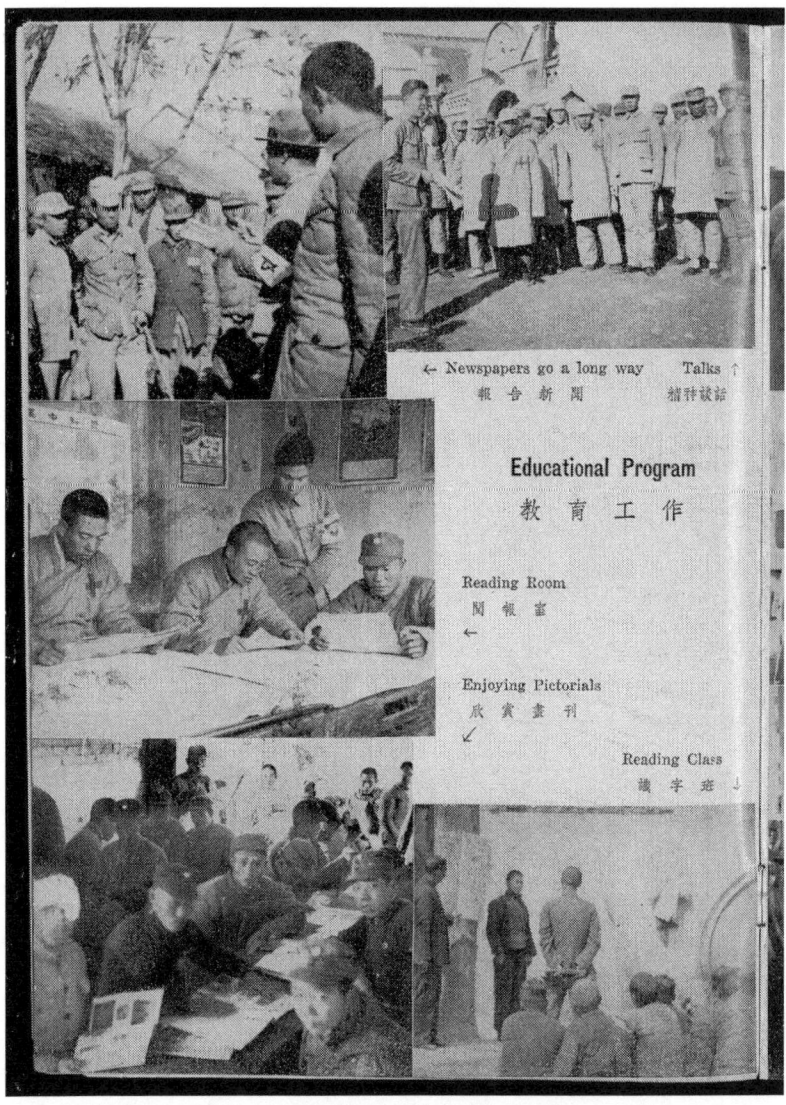

← Newspapers go a long way　　Talks ↑
報　告　新　聞　　　　檜　升　談　話

Educational Program
教　育　工　作

Reading Room
閱　報　室
←

Enjoying Pictorials
欣　賞　畫　刊
↙

Reading Class
識　字　班 ↓

Chinese YMCA workers providing educational programs for soldiers, such as reading newspapers to soldiers, setting up a reading room where soldiers could read pictorials, and offering reading classes for illiterate soldiers.

Y Huts and Club Houses
青年會軍人招待處及俱樂部

Such Huts are built along the Main Lines of communications and are subjected to constant bombings. In many places they have been destroyed by bombs four or five times in one month but to be erected again to meet the need.

此種招待處通設於交通要點，而屢�轟炸，每月數次，但立卽重建，以應急需。

Temples and other public establishments are also good club houses.

廟宇，祠堂，及其他公共處所，均成最好之軍人俱樂部。

The Club for Wounded Soldiers of the Chinese YMCA, where professionals provided medical, entertainment, and educational services for wounded soldiers.

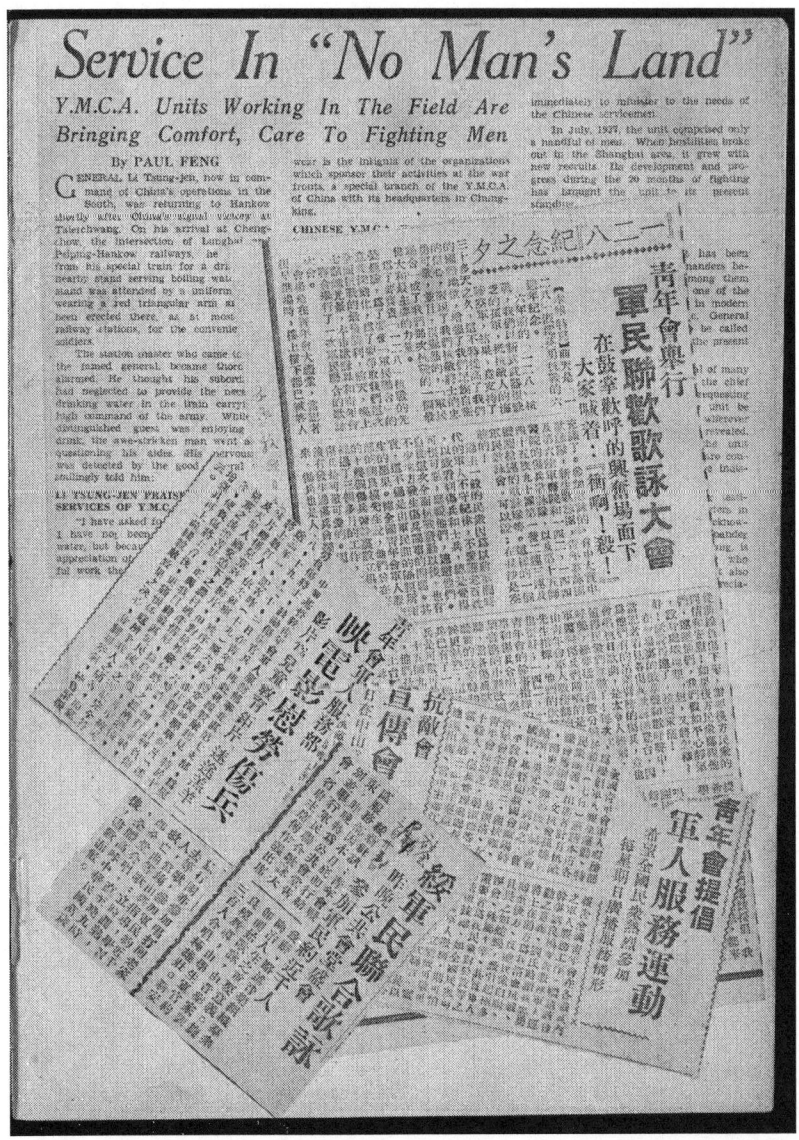

Newspaper clippings reporting that the Chinese YMCA served soldiers by showing movies for them and organizing singing assemblies where they participated with the civilians.

This 1942 US propaganda photograph for encouraging support of America's World War II allies shows a smiling Chinese soldier with the caption "This man is your friend. He fights for freedom." (Courtesy of Special Collections, Hennepin County Library, Minneapolis, Minnesota, Local identifier: mpw00292)

Generalissimo Chiang Kai-shek inspecting student soldiers enlisted in the Chinese Youth Army, 1944. (Courtesy National Archives, photo no. 111-SC-208797)

5

The Construction of the Soldier Ideal by Educated Youths

This book has examined several different images of Chinese soldiers depicted in literary works. Similar to the literary images of Chinese soldiers, the topic of student soldiers during the Second Sino-Japanese War has also been largely forgotten in Chinese military history scholarship. This chapter discusses how educated youths constructed the soldier ideal to fit their concerns.

As the Western allies entered the war against the Japanese after the attack on Pearl Harbor in December 1941, China's resistance war against the Japanese became an important theater of the Second World War (1939–1945). The new domestic and international situation forced Chiang Kai-shek to adjust his military mobilization policies. Chiang eventually launched the Campaign to Mobilize Educated Youths to Join the Army in late 1944, calling upon educated youths, mainly high school and college students, to join the army. Conscripting educated soldiers had been part of the GMD's continued efforts to forge a militarized citizenry following the compulsory conscription law in the mid-1930s. Now, the GMD government elevated the political status of student soldiers in order to attract them to the army. These students viewed performing military service as a way to get involved in national politics and to extend the principles of self-government and democracy into the Youth Army. This chapter first discusses the Campaign to Mobilize Educated Youths to Join the Army to provide a historical background for the GMD's efforts to militarize educated youths. Then it analyzes the writings by the student soldiers to examine what soldier ideal they advocated and practiced in their army life.

The Campaign to Mobilize Educated Youths
to Join the Army

The compulsory conscription law issued by the Chinese Nationalist state in 1933 stipulated that performing military service was the national duty of males in the Republic of China. However, the 1936 revision of the compulsory conscription law made it clear that students and graduates of the high school level and above were exempted from military duty, together with those who did not meet physical standards and who were only sons in their families. Historian Jay Taylor notes, "Seven years into the Second Sino-Japanese War, the Chinese government had not conscripted college students or graduates with the exception of some doctors, engineers, and English majors, the last to serve as interpreters."[1] There were several reasons for this abstention. One would be the Chinese tradition that the army was made up of the lower classes, as well as the generally poor treatment of the rank and file of the Chinese Nationalist army, which discouraged student military participation. Another factor was the very small number of people in Republican China who had received higher education. According to the memoir of the GMD official Chen Lifu (1900–1993), who served as the wartime minister of education, when the full-scale Second Sino-Japanese War broke out in 1937, "college-educated people numbered only 40,000 in the entire nation. Not even one of every 10,000 persons had attended college. Our nation's future depended on these young educated people."[2]

China's exemption of students from military service was unexceptional. Japanese students were exempted from conscription during the Meiji Restoration. The Meiji government enacted its first conscription law in 1873. It stipulated military duty for males, but did not conscript students who were at or graduated from the secondary or higher educational institutes. The Meiji government, at the suggestion of a German military advisor, issued an amendment to the conscription law in 1889. This amendment adopted a volunteer system modeled on the German style of training, which offered one year of special training for volunteers who had certain academic backgrounds and then appointed them reserve officers. As the Japanese scholar Mitsutoshi Hanyu maintains, this step aimed to attract men of the upper class into the military, especially the educated and the rich.[3] After World War I, Japan took further steps to cultivate military training among students. "Preparatory military drill became compulsory for students under the conscription age, and active-duty officers were assigned to junior and high schools for this training."[4] Students who had passed the preparatory military drill and graduated from vocational school or an equivalent educational level could be appointed as a reserve second lieutenant after eleven months of training. Although students were still legally exempted from military service, they were considered potential army trainees and reserve officers before World War II.

Having studied in Japan at a preparatory school for the Imperial Japanese Army and then served in that army from 1909 to 1911, Chiang Kai-shek was aware of the trends in military training for youth in Japan. He had been advocating the significance of military training for students since the early years of the Second Sino-Japanese War. In his wartime diaries that recently became available to the public, Chiang frequently wrote that the GMD state should place as its priority the organization and training of student soldiers and boy scouts among elementary, middle, and senior schools. Chiang often did research on how other countries organized and trained their youth.[5] He commented that the educational system should be grounded on military discipline and spirit, and that students in elementary, middle, and senior schools and colleges should all receive military training.[6] In his efforts to militarize the society, Chiang also advocated that military virtues, such as following military discipline and order, obedience to authorities, and bravery, should be indispensable to the militarized citizenry ideal.[7] The frequent stress in Chiang Kai-shek's diaries on the importance of mobilizing students to receive military education and to cultivate military virtues reveals his intention of extending the militarized and disciplined citizen-soldier ideal to the group of students.

Despite Chiang Kai-shek's intention of military mobilization among students, the Campaign to Mobilize Educated Youths to Join the Army was not launched by the GMD government until October 1944. This campaign was launched to meet a series of internal and external challenges as the Second Sino-Japanese War entered its last phase.

Domestically, the major crisis was severe difficulties with conscription by the later years of the war. The poor training of new conscripts coupled with the hampered implementation of the conscription system made consistent military training for a growing military force difficult to continue. According to the recollections of Nationalist army general Bai Chongxi (1893–1966), the new conscripts had to join the front line before they had completed the necessary military training and fully familiarized themselves with using weapons. Because of these problems, the Nationalist army's combat effectiveness severely deteriorated.[8] Chiang Kai-shek delivered a speech on behalf of the Military Affairs Committee on January 10, 1944, which said: "Because of the long-established social discrimination against the soldiers, the majority of the nationals in our country have not fully embraced the concept of performing military service in accordance with the conscription law. Many able-bodied men claim exemption from performing military duty by pretending that they are students studying at high schools. The inability of the army to attract the participation of educated youths has severely reduced its fighting capacity. This is the biggest mistake of our army."[9]

The reality of the reduced combat capacity of the Nationalist army during the last few years of the Second Sino-Japanese War has been discussed by the historian Hans van de Ven. He points out that until 1941 the Nationalists were relatively successful, as they drew nearly 2 million men each year into the army while maintaining social stability. However, because of the embargo imposed by Japan, China faced financial chaos, and after 1941 the combat capability of the Nationalist forces dropped dramatically. During this period the pool of men who could be recruited without serious consequences for local agricultural productivity had dried up. Another problem was that "large parts of the Nationalist army became militarily useless and corrupt, engaging in smuggling, and trading with the Japanese."[10]

Driven by the decreased capability of combat troops within the army, the GMD government hastily designed new policies to reform the military. It reduced the size of the army, demanded that the army supply its own needs, and tolerated the army's participation in smuggling and other nefarious activities.[11] In 1944, Chiang Ching-kuo (1910–1988) made a proposal to his father, Chiang Kai-shek, arguing that "what China desperately required was a new sort of military force composed of literate and patriotic youths led by officers of high quality and dedication."[12] This proposal to organize an elite army was accepted by Chiang Kai-shek.

Chiang's acceptance of the proposal was a response not only to the conscription crisis at home but also to the international situation of China's resistance war. The Pacific War started on December 7, 1941, with the Japanese attack on the US naval base at Pearl Harbor. American bases at Guam and Wake Island as well as Hong Kong were lost in the same month. After being driven out of Malaya, British forces attempted to resist the Japanese during the Battle of Singapore but ultimately were forced to surrender on February 15, 1942. The rapid collapse of Allied resistance drove twenty-six countries, including China, Britain, the United States, and the Soviet Union, to meet in Washington, D.C., on January 1, 1942, where they signed a manifesto of war against the Axis countries of Germany, Japan, and Italy, vowing that no one would make a separate peace with the enemy.

As a result, the Second Sino-Japanese War was folded into the Asian theater of World War II. On January 3, 1942, Chiang Kai-shek was named commander-in-chief of the Allied units in the China war zone, responsible for commanding the Allied forces in China, Vietnam, Burma (present-day Myanmar), and Thailand.[13] In February, he set up the Chinese Expeditionary Army (*Zhongguo yuanzhengjun*) of over one hundred thousand soldiers, and it entered Burma to fight alongside the British Army for the first time. This unit was mainly armed with US equipment and was subject to training by US military officers.[14] The 1st American Volunteer Group, better known as the "Flying Tigers," had begun training in August 1941 in Burma to defend China against the Japanese. Led by Major General Claire Lee Chennault (1893–1958), they were formally incorporated into the US Army as the 14th Air Force Battalion in July 1942 with a new base in Kunming, Yunnan. Chiang continued to receive supplies and weapons from the United States as the Chinese conflict merged into the Asian theater. By 1944, the Americans had implicitly agreed to create a modern Chinese air force[15] and committed to training and arming thirty-nine Chinese army divisions.[16] As a result, an army consisting of well-educated soldiers was needed to master the newly introduced foreign weapons and to communicate with Allied forces.

To meet the domestic crisis and international challenge, the students' exemption from performing military duty was cancelled in the revised conscription law issued in March 1943. Chen Cheng (1897–1945), who was appointed the commander of the Chinese Expeditionary Army in 1943, requested that the Military Affairs Committee enlarge the army by recruiting well-educated soldiers.[17] The committee demanded that every city in Sichuan Province select three hundred educated youths to join the expeditionary army. By the end of 1943, 27,129 students in the province had volunteered.[18] The campaign gradually developed in another ten provinces.[19]

However, the official nationwide military conscription campaign among students was not launched until October 1944. The immediate reason for this campaign was that China's national resistance efforts at home faced a severe challenge when the Japanese forces launched Operation

Ichigō in April of that year. The Nationalist army lost almost 600,000 soldiers and most areas in Henan, Hunan, Guangxi, Guangdong, Fujian, and Guizhou Provinces. In Autumn 1944, the Japanese attacked Guangxi and Guizhou, posing serious threats to the GMD's wartime capital of Chongqing. Minister of National Defense He Yingqin wrote in his memoir: "In early September of 1944, 150,000 Japanese enemies invaded Guilin and Liuzhou, but our available units had less than 120,000 soldiers and suffered from difficulties in transportation; both factors caused our loss of the battle."[20] The setback in resistance efforts at home forced the GMD government to adjust its military policy and work on mobilizing the students to constitute an elite army.

On October 11, 1944, the GMD government held a conference in Chongqing in an attempt to mobilize the educated youth across the nation to perform military duty. It established the General Supervision Office for Conscripting and Training the Educated Youth Army in the Military Affairs Commission (*Junshi weiyuanhui quanguo zhishi qingnian zhiyuan congjun bianlian zongjianbu*) to handle the conscription and training of educated youths. Many military and political authorities as well as college presidents served as members of this committee.[21] On October 24, 1944, Chiang Kai-shek delivered a speech calling upon educated youths throughout the country, including high school and college students, as well as teachers and civil servants, to enlist in the Youth Army. According to this speech, educated youths between the ages of eighteen and thirty-five who had graduated from high school or above, as long as they met physical standards, were encouraged to join the army.[22] Chiang pointed out that although some educated youths had participated in military-related work or joined the military academy, the performance of the educated youths in military service had not been popularized at the national level. Chiang also recognized the patriotic spirits of the students and claimed this exemption in the conscription law was a hurdle to their patriotic action.[23] As Chiang later said concerning the speech, "We had reached the critical point which could determine the result of the war and the destiny of our nation. The following year would be the crucial period for us to achieve the ultimate victory. This was surely a golden opportunity for our educated youths to serve the nation."[24] In November 1944, Chiang Kai-shek named Minister of War Chen Cheng to command the officially named Educated Youth Expeditionary Army (*Zhishi qingnian yuanzhengjun*).

Launching the Campaign to Mobilize Educated Youths to Join the

Army shows that the GMD government strengthened its efforts to broaden the militarized citizen ideal to include educated youths and further militarize the society. In his speech "Gao zhishi qingnian congjun shu," Chiang expressed his expectation for educated youths to fight as real combatants and perform a frontal attack on the enemy (*chongfeng xianzhen*) as a revolutionary vanguard (*geming xianfeng*).[25] Chiang also expressed his belief that enlisting educated youths into the army could "fundamentally overthrow the social discrimination against the military profession."[26] The national crisis was ascribed to "the intellectuals' value of putting intellectual pursuits above martial arts" (*zhongwen qingwu*).[27] Chiang encouraged not only high school and university students but also school teachers and professors to volunteer to join the army.[28] He advocated that their participation in military service would extinguish the discrimination against martial arts in the society.[29]

To achieve militarization among the students, the GMD had to develop strategies to make military careers attractive to them. Chang Rui-te notes that because of the flowering of Chinese nationalism in the 1930s, a career in the military enjoyed unprecedented popularity among the youth during that period. Many students indicated their desire to pursue a military career, and the prestige of becoming a military officer ranked higher than becoming a doctor or lawyer. However, by the time the Second Sino-Japanese War had entered its middle stage after a long succession of embarrassing defeats, the pursuit of an army career had lost its appeal for most Chinese youths. As noted in chapter 2, a decline in military pay and benefits also added to the problem.[30] Since a military career was not popular among educated youths in the later years of the war, Chiang had to enhance the significance and status of the Youth Army in order to militarize the students.

One way Chiang Kai-shek enhanced the significance of the Youth Army was to celebrate them as "the model and backbone for all the military units in the nation to promote their fighting spirit and improve their combat capacity."[31] In the 1944 speech "Qingnian yuanzhengjun bianxunde yaozhi" (The key principles in training the Educated Youth Expeditionary Army), Chiang claimed that the participation of educated youths could raise the intellectual level of national armies and their ability to employ superior weapons and advanced military tactics to the utmost.[32] Chiang stated that educated youths were intelligent and cultivated enough to judge the enemy's situation and to fight independently.[33] This would allow one soldier in the Youth Army to defeat ten of the enemy; and the strength of one division of the Youth Army would be equivalent to that of ten divisions of common soldiers.[34] Chiang's elevating the significance of the Youth Army as a model and backbone for the national army revealed his intention that the Youth Army, after proper training, could raise the quality of the Chinese military in the war.

Another type of rhetoric Chiang often employed to make military careers more appealing to educated youths was his advocating that military knowledge and experience in combat were an indispensable part of the youths' study and development. In "Dui congjun xuesheng xunhua" (A lecture to student soldiers), delivered on January 10, 1944, Chiang claimed: "You [the student soldier] could enhance your knowledge, skills and physical capacity by performing military training; this experience could not be achieved from any regular school."[35] When Chiang attended the Chongqing conference on October 11, 1944, he discussed his own army experience in an attempt to illustrate the significance of army service for the personal development of educated youths: "With forty years of experience in the army, I feel that joining the army is not only a correct path to defend the nation. The knowledge and skills I have acquired from the army experience are the most precious scholarship to me. The structure of the army is the most advanced one among all social organizations. Possessing military knowledge and skills and understanding the operation of the army structure is extremely useful for one's career development. Military service is not only a joyful experience but also the most important education in one's life."[36] Chiang shared his personal army experience to demonstrate that performing military service was not only significant for the nation but also beneficial for the personal development of the individual soldier.

The importance of military service for the personal development of educated youths was also stressed in Chiang's other speeches. In the speech "Gao zhishi qingnian congjun shu" (A speech to call on educated youths to join the army), Chiang claimed that "the battlefield was the only school for the youth to build the foundation for their career," and that "the youth could not achieve their personal ambition without army experiences."[37] According to Chiang, the army could give youth useful and practical knowledge and skills, promote their physical strength, and help them develop a deep understanding of life; none of the above could be learned from any regular school.[38] In "Qingnian yuanzhengjun bianlian zhi tezhi yu jiaoyu yaoxiang" (The peculiarity and key points in training the Educated Youth Expeditionary Army), delivered to the Youth Army officer trainees on January 16, 1945, Chiang first praised young students as being well educated academically and fully indoctrinated with the beliefs of the Three Principles of the People.[39] He celebrated them as "the elite of the society and the treasure of the nation."[40] However, Chiang pointed out that in order to become competent cadres in the postwar reconstruction, they needed to acquire military knowledge and combat experience.[41] Chiang used this logic to justify his intention of extending the militarized citizen ideal beyond the war effort.

In order to further justify the importance of army experience for the development of the educated youth, Chiang proclaimed that the Youth Army was "the best professional school" for educated youths because it taught them practical skills, thereby increasing their ability to lead the ranks.[42] According to Chiang, "The education and training in the army was not only the foundation for academic studies in regular schools but also more practical than regular school teaching."[43] The Youth Army encouraged every student soldier to identify one skill that was necessary for daily life to learn in his spare time. With a strong physical body, substantial knowledge, superior skills, and a deep understanding of the operation of the army structure, the student soldiers could serve as the social backbone and be capable of leading the masses after the war.[44] By advocating the Youth Army as the best professional school, Chiang attempted to make military service attractive for educated youths in order to persuade them to join the military.

Having stressed the importance of army experiences for the personal development of educated youths, Chiang encouraged them to further contribute their intellectual expertise to the army. Within one month after they were recruited to the army, educated youths were required to complete a registration form providing information on their interests, expertise, levels of scientific knowledge, and technological skills. Based on this information, the Youth Army appointed soldiers to positions that matched their individual skills. For example, around ten thousand vehicle drivers and radio communication personnel were recruited from the Youth Army.[45] Encouraging the educated youths to apply their expertise to the military was made clear by Lieutenant General Luo Zhuoying (1896–1961), who served as commandant of the Army Officers Training Center between 1944 and 1945. He told a *Dagongbao* reporter on November 14, 1944, that the Youth Army training was designed based on subjects each individual soldier had learned before he joined the army. For instance, a student soldier who had studied civil engineering would be trained for performing projects such as bridge and road construction, whereas a student who studied physics would be trained to examine the composition and usage of new weapons; a student soldier who had studied chemistry would be trained for producing bombs. Luo used these detailed examples to illustrate how military training advanced rather than impeded the students' academic pursuits.[46]

To further demonstrate the compatibility of military training with personal development, Chiang Kai-shek argued that training in the Youth Army needed to combine military discipline and techniques with civic skills and artistic knowledge. Chiang borrowed the Confucian conception of *liuyi* (six arts) to support this training philosophy. According to the *Zhouli* (Rites of Zhou), Confucius (551–479 BCE) considered six areas, named *li* (rites), *yue* (music), *she* (archery), *yu* (charioting or horsemanship), *shū* (literary writing and calligraphy), and *shù* (mathematics), to be essential for the education of the *junzi* (superior man or perfect gentleman) as a leader in civil-military affairs. Chiang explained *li* as military discipline, *yue* as artistic genres, such as music, dance, and fine arts, *she* as shooting, *yu* as driving military vehicles, *shū* as writing military orders, drafting work reports, and recording statistics, and *shù* as mathematic knowledge.[47] Chiang claimed that a combination of martial techniques, civic studies, and artistic knowledge should be the educational foundation for modern nationals, including Youth Army soldiers. Chiang appropriated *liuyi* to advocate applying the militarized citizen ideal to the youths' education.

After educated youths were recruited and assigned to their respective divisions, they were subjected to a two-week basic training course that was intended to familiarize them with military discipline. The training included politicizing through ideological indoctrination and instilling loyalty to the national leadership. Its main focus was regulating the daily practice of educated youths in order to turn them into soldiers. In the speech "Qingnian yuanzhengjun bianxunde yaozhi" (The key principles in training the Educated Youth Expeditionary Army), delivered on December 7, 1944, Chiang told Youth Army officers to pay special attention to regulating educated youths' behaviors in daily life.[48] Chiang said: "The first step to train educated youths is to help them form proper life habits by guiding them on details of their daily behaviors, such as dress, eating, sleeping, as well as grooming, doing laundry, washing shoes, cutting fingernails, etc."[49] This two-week basic training demonstrated Chiang's attempt to politicize and discipline educated youths upon their recruitment into the army.

By launching the Campaign to Mobilize Educated Youths to Join the Army in late 1944, the GMD government tried to extend its politicized, militarized, and disciplined citizen ideals to high school and university students as well as civil servants—all groups that were exempted from compulsory military service prior to 1943. To justify the military mobilization campaign among educated youths, Chiang employed a series of strategies to make the military profession attractive to them. He exalted the status of the Youth Army as the model and backbone for national armies while advocating that military knowledge and experience were indispensable to the youths' development, with the Youth Army being the best professional school. In terms of training principles, Chiang encouraged educated youths to apply their expertise to the military in order to advance their studies. With this in mind, in order to attract students with different majors to join the army, he also argued for national education that involved a combination of military techniques, civic studies, and artistic knowledge.

The Self-Government Ideal in the Youth Army

The Campaign to Mobilize Educated Youths to Join the Army was largely embraced among the youth in China. By the end of 1944 the Youth Army had drafted approximately 140,000 students across the nation.[50] According to a report written by a student soldier named Zhang Jingcang in January 1945,[51] 55 percent of the Youth Army soldiers in his one-hundred-man company were high school students, 30 percent were university students, and the rest were civil servants, journalists, and other young professionals.[52] After the recruited youths started military training, they wrote diaries and reportages in their free time, most of which were published in the literary supplement of the GMD's official newspaper *Zhongyang ribao* (Central daily news).[53] The fact that the GMD published these writings in its official newspaper implies its intention to propagandize its military mobilization campaign to the public and justify its militarized ideal of citizenship.

In their writings, student soldiers in the Youth Army supported and welcomed the GMD's effort of militarizing educated youths. Zhang Qing argued in the article "Junzhong wenxue" (Literature and the army), published on April 26, 1945, that literature and the army shared the same spirits and rules. The army's operation relied on a strict hierarchy, and literary works also required a compact structure. The army advocated a simple lifestyle, while literary writers utilized unsophisticated words to illuminate profound meanings. Military training demanded solid exercise, and literary writings also required hard practice.[54] Hou Bingchen reported on February 3, 1945, in "Xin jun xin shenghuo" (New army and new life) that many Youth Army soldiers preferred army life because there were no examinations or trivial bureaucratic work. They felt that life in the army purified their minds since they did not delve into romantic affairs or deal with meaningless social engagements.[55] In the 1945 report "Laodong yu yule" (Labor and entertainment), Zha Mi explained his belief that transforming biased attitudes of society toward soldiers required the mutual efforts of the government and educated youths.[56]

Youth Army soldiers also embraced the significance of military training for their personal studies, as advocated by Chiang Kai-shek. They treated the army experience as a way to enhance their knowledge and improve their physique. Lian Lulu wrote in the 1945 report "Wanhui" (An evening party) that Youth Army soldiers believed if a young person were able to both fight in combat and produce literary works, then he was a true master of both military and civilian scholarship.[57] In the 1945 report "Jiyu fulao" (To the elders in the family), Zhao Tingjun wrote: "After the two-year tough training in the army and the ordeal of the war, I would grow into a steadfast and resolute person."[58] Luo Qiqian wrote in his 1945 report "Qingnianjun yu qingnian Zhongguo" (The Youth Army and young China) that military training would prepare him to be a social leader by strengthening his body and enhancing his knowledge.[59] A year earlier, Jia Yunfu's report "Qingnian lushang" (On the road being built by the youth) described how the Youth Army soldiers happily worked on building roads as part of their military training. They competed to work harder even though their palms were excoriated by hard labor.[60] Zha Mi wrote in the 1945 report "Laodong yu yule" that Youth Army soldiers used their tender hands to dredge mud. Zha commented: "If the students and civil servants did not join the army, they would never have the courage and knowledge to perform labor work."[61] Zhao Tingjun wrote in the 1945 report "Jiyu fulao" that the student soldiers were getting stronger and more vigorous because of their regular life schedule and tough military training.[62] Zheng Bingsen wrote in his 1945 public letter to mothers of the Youth Army soldiers that he could stand for hours and walk many miles carrying heavy materials without feeling exhausted. He and his fellow soldiers managed to complete the construction and decoration of their barracks in one day. He was also brave enough to walk in the dark and not be afraid of rats. He had learned how to hide himself from the enemy, and how to search for and shoot targets.[63] The Youth Army soldiers reported in their writings that, thanks to military training, they improved their health and physique, grew braver, and learned a substantial amount of new knowledge. They expressed pride in these changes brought by their military experiences.

Although student soldiers cooperated with the GMD's military mobilization by joining the Youth Army, they still managed some form of independence and expressed their concerns about building an army based on the self-government principle. The self-government principle was not a new idea invented by student soldiers. The argument for educational independence had been advocated by overseas Chinese students and professional educators in China since the early years of the Republican era. It was a product of the emergence of local activism in China in the 1920s. In 1902, twenty-three Chinese students in California formed the Chinese Students' Alliance, which became a nationwide organization of Chinese students in the United States. This alliance was active until 1931 and allowed overseas Chinese students to make acquaintances and exchange ideas with their fellow countrymen. As Stacey Bieler comments, the Chinese Students' Alliance was a laboratory of self-government, a nonscholastic activity that would help the students develop into the American ideal—an independent, well-rounded person.[64] This alliance published *Chinese Students' Monthly* as their official magazine, which became a most influential periodical among Chinese students in the United States at that time. The magazine discussed domestic and international political events and important social movements in China in the early twentieth century, among which student self-government in education was a central theme. In an article titled "Christian Education in China" published in this magazine in 1923, Chen Lang Tung argued that Christian schools had been centers for radiation of the spirit of democracy in China because they supported student self-government, relations between the students and the faculty, and freedom of discussion and expression of opinions in classrooms and various student organizations.[65]

The advocacy of student self-government by overseas Chinese students in the 1920s was accompanied by the phenomenon that students had become a new force of political movements since the 1919 May Fourth

Movement in China. In 1921, the National Federation of Education Associations, established in Tianjin in 1915, drafted a new educational system, which was accepted by the Ministry of Education.[66] This new system, copied from American models, aimed at educational independence from political bureaucrats. Through this system, the NFEA promoted student self-government in the schools and assisted in the running of local student societies. These experiences "gave students an elevated sense of their importance in school and national affairs."[67] As Kristen Mulready-Stone shows in her study on Shanghai youth, the youth emerged as a distinct segment of society and gained great prominence in Chinese national affairs beginning in the early twentieth century.[68] Mobilizing young students into the military showed that the Chinese Nationalists recognized the importance of involving the youth in their resistance and state-building efforts.

Although the students collaborated with the Nationalists in the resistance efforts by joining the army, they treated the Youth Army as a laboratory to practice self-government. The writings by the student soldiers reveal how they applied this principle in their study and entertainment life. The soldier Zhao Tingjun wrote in his 1945 "Jiyu fulao" that every company had a club called *Zhongzheng shi* (Sun-Yat-sen room), where the soldiers played chess and musical instruments and organized a party once a week.[69] Qiu Hongyi wrote in his report "Women zai Hufeng" (We are in Hufeng) on February 5, 1945, that every evening the Youth Army soldiers read books, wrote diaries and letters for one hour, and at least once every week gathered in small groups to exchange their thoughts and works.[70] Chen Can wrote in "Womende xin feidao zhanchang" (Our hearts have flown to the battlefield) on February 10, 1945, that the soldiers organized their own soccer teams, singing teams, and wall poster teams.[71] According to Ouyang Wenhui's report "Zhenggongban shenghuo zhuishu" (Memories of the life in the class for political workers), written in January 1945, the Youth Army soldiers organized an exhibition to show each other their art works.[72] These writings show that the Youth Army soldiers actively participated in building the Youth Army by practicing self-government in their entertainment and study.

The Youth Army soldiers also practiced the principle of self-government in the army administration. According to Zhang Jingcang's report "Xinzhanshi xinzuofeng" (New soldiers and new styles), published on January 29, 1945, before military instructors arrived and formal military training began, the Youth Army soldiers took the initiative in establishing self-government councils. These councils set rules to regulate the behaviors of the Youth Army soldiers and patrolled the area around the military base, preventing soldiers from escaping.[73] According to Zhang, the efforts of the Youth Army soldiers to govern themselves even received the support of the officers because it made the officers' jobs easier.[74] Shen Yi recorded in "Kaishile junying shenghuo" (The army life has started), published on January 31, 1945, that every company had a self-government council, which had several branches, including order-maintaining (*jiucha*), entertainment (*kangle*), research (*yanjiu*), and general affairs (*zongwu*). The council members were elected by the Youth Army soldiers themselves, and they judged the behaviors of the soldiers and carried out punishments if any soldiers violated the established rules.[75] Zhang Zhengquan's report "Junshiduide yingzhong shenghuo" (The army life in the base), published on February 10, 1945, explained that the self-government councils regulated every aspect of life and training in the Youth Army, including diet, sanitation, and event organization.[76] According to Ouyang Wenhui's "Zhenggongban shenghuo zhuishu," when the guests gave a lecture to the Youth Army soldiers, the self-government council members prepared the lecture room and provided service for guests. The self-government councils were also in charge of receiving correspondences and distributing them to each soldier.[77] The sources examined above show that the student soldiers celebrated the soldier ideal through self-government and active participation in army administration.

Their practice of self-government encouraged the Youth Army soldiers to advocate for an equal officer-soldier relationship and a democratic spirit in the army. In the report "Xin jun xin shenghuo," Hou Bingchen recorded that when the Youth Army soldiers felt their treatment by the company commander was unfair, they organized a protest against the commander. Some soldiers even directly confronted the commander. Their protest usually resulted in the commander's apology. For instance, the commander let the soldiers decide on the deadline for when to return to the base before they were released for holidays. According to Hou, this concept of democratic spirit also applied to financial expenditures, as they were completely transparent and were managed by the soldiers themselves. The company commander filled the role of the military leader, but not that of the financial minister.[78] According to Qiu Hongyi's report, "Women zai Hufeng," written in 1945, the company commander and the soldiers ate and slept together, and the army maintained the principles of freedom, equality, and fairness and did not have unjustified punishment.[79] Some anonymous soldiers reported in "Chufa xingjun ruying" (Marching to the military base), published on January 15, 1924, that there was no estrangement between the army officers and the soldiers.[80] These writings show that the Youth Army soldiers actively pursued a democratic administration within the army. Their pursuit made a difference in the army culture by forcing lower-level army officers who directly led the Youth Army soldiers to compromise and support them.

The ideals of self-government and democracy encouraged educated youths to voice their dissatisfaction with the political and military authorities in their writings. In his 1945 report "Kaishile junying shenghuo," Shen Yi pointed out the discontent felt by the Youth Army soldiers toward the government. Shen wrote that educated youths joined the army so as to achieve democracy. He requested that the army improve their training techniques by promoting self-government and the democratic spirit among the soldiers.[81] The report "Chufa xingjun ruying" argued that "the best strategy to train us [the Youth Army soldiers] is to stimulate our initiative," and that "delivering political sermons to educated youths was like displaying accomplishments in front of the experts (*banmen nongfu*)."[82] The Youth Army soldiers publicly asserted that their biggest concern was the realization of democracy, and they expressed their eagerness to learn practical knowledge, such as how to use new weapons and drive vehicles, rather than spending time listening to political sermons.[83] By advocating in the GMD's official newspaper a soldier ideal that valued active and direct political participation in administering the army, the Youth Army soldiers confronted the GMD's discourse that celebrated a highly politicized and disciplined soldier ideal.

This chapter focuses on the Campaign to Mobilize Educated Youths to Join the Army that was launched by the GMD government in late 1944. It examines the historical background of the campaign, the rhetoric Chiang Kai-shek used to appeal to educated youths, and the writings by the Youth Army soldiers published in the GMD's official newspaper. It shows that the Nationalist state strengthened its efforts to militarize society in the later years of the Second Sino-Japanese War. The student soldiers actively participated in the GMD's state-building effort to build a strong army by joining the Youth Army during the conscription campaign. However, an examination of their writings shows that their alliance with the GMD government was ambiguous. Although they supported the Nationalist rhetoric of militarizing educated youths, they asserted their soldier identity based on the principles of self-government and active and direct political involvement. In doing so, they challenged the GMD's disciplined and politicized citizen-soldier ideal.

Chiang Kai-shek launched the nationwide military campaign among the youth during the period when China's resistance war reached a crucial period. A combination of domestic and international situations, including the deteriorations of the conscription system and the combat capacity of the Nationalist army, the merging of China's resistance war into the Pacific War and the Second World War, the commencement of Operation Ichigō, and China's increasing dependence on the Allies, forced Chiang to think of new initiatives to increase the capacity of the Nationalist army and improve its international image. In this context, Chiang made efforts to militarize educated youths, as he believed that they were capable of learning how to use new weapons quickly, communicating with the Allied forces smoothly, and making judgments on the enemy's situation independently.

Mobilizing educated youths into the army was not a sharp shift from the previous policy of exempting students from military duty. It revealed Chiang Kai-shek's long-term state-building intention of extending the politicized and militarized citizen ideal to the society. Chiang employed a series of rhetoric-based strategies in order to justify his plan to educated youths and to make military service appealing to them. He elevated the status of the Youth Army as the model for national armies, and he advocated the significance of military knowledge and experience for the youths' development by claiming the Youth Army was the best professional school. Chiang also argued that applying their expertise to the military could help educated youths advance their studies.

The soldiers in the Youth Army embraced the GMD's state-building effort of militarizing educated youths. The writings by the Youth Army soldiers show that they accepted military training as an exercise to improve their physical abilities and learn practical knowledge and labor skills. However, they argued for a soldier ideal based on active political involvement alongside their military development. In the writings that were published in the GMD's official newspaper, educated youths who joined the Youth Army identified themselves with the role of national spokesmen when they expressed their pursuit for self-government and democracy. They believed that the officer-soldier relationship should not be regulated by a rigid hierarchy; instead, it should be based on the principle of democracy and self-government. Educated youths who joined the Youth Army treated the army as a laboratory to practice these principles.

The GMD government and educated youths in the Youth Army each constructed their own discourses of the soldier to assert their goals and visions of how to build a strong army. Another discourse of the soldier that coexisted with the GMD's soldier ideal was the one forged by the CCP in the revolutionary base of Yan'an, which will be the focus of next chapter.

6

The Army-People Bond in Mass Culture in Wartime Yan'an

The previous five chapters have shown that political, social, and cultural forces in the GMD areas constructed discourses of the soldier figure to meet their state-building goals and to articulate their respective political influences. Another influential political force in modern China was the Chinese Communists in their revolutionary bases during the Second Sino-Japanese War. This chapter examines how the CCP in Yan'an constructed the soldier figure in its social movements and mass culture during the war. This case study demonstrates that the CCP constructed the soldier figure along the line of an emotional bond between the army and the people. This rhetoric was essential for the CCP's state-building agendas of winning support from the peasants in its areas and forging closer social integration in its revolutionary base.

During the Second Sino-Japanese War, the CCP accumulated strength incrementally, village by village. Yan'an, a northwestern inland city isolated from eastern urban centers, became the seat of the Chinese Communist government of what became known as the Shaanxi-Gansu-Ningxia Border Region (Shaan-Gan-Ning *bianqu*, 1937–1947).[1] In the first year of the war, the total Communist military strength was about ninety-two thousand, but by the end of the conflict the Communist military forces had grown to 1 million regular troops, augmented by 2 million militia.[2] To ensure continued popular support for military recruitment, the Chinese Communists made strenuous efforts to build army-civilian solidarity in Yan'an and other revolutionary bases.

Substantial cultural productions helped build the image of the soldier figure and army-people solidarity. Popular slogans such as "the army and the people are as close as one family" (*junmin yijia qin*) and "the army and

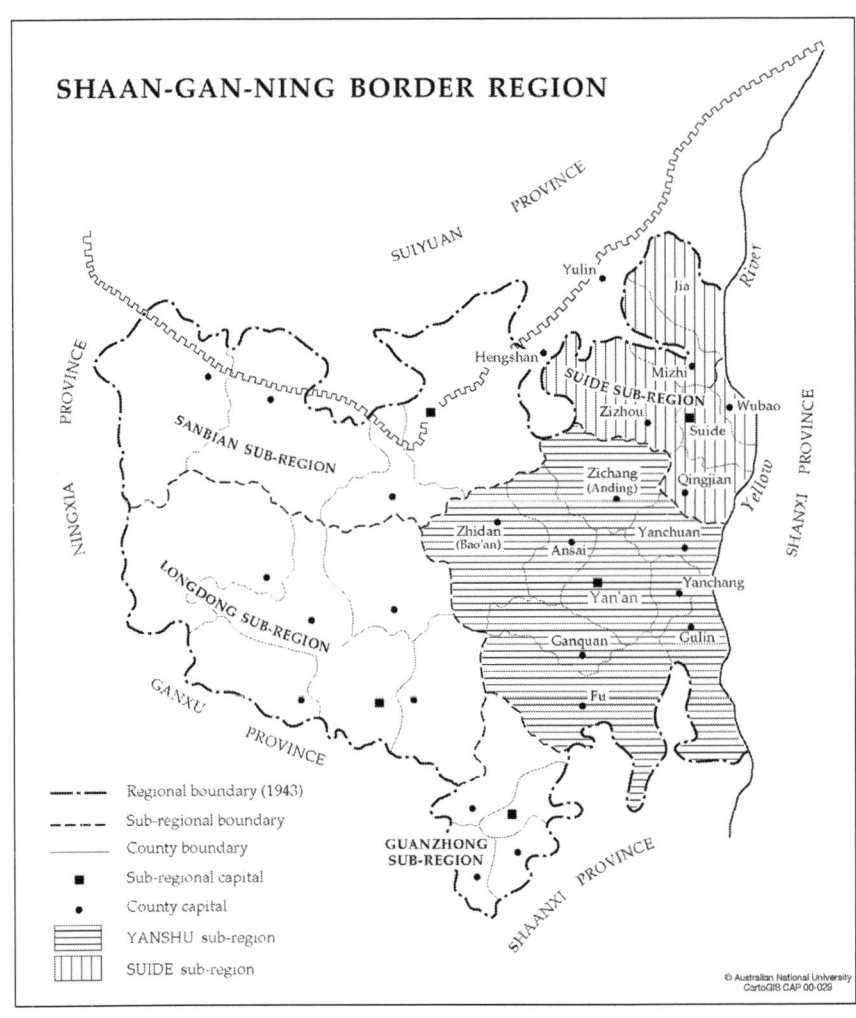

The Chinese Communist Shaan-Gan-Ning Border Region. (Courtesy of Cargo-GIS Services, College of Asia and the Pacific, The Australia National University)

the people bond like fish and water" (*junmin yushui qing*) dominated mass culture, which was facilitated and appropriated by the Communists as a propaganda medium in Yan'an. The soldier figure fervently celebrated in Yan'an's mass culture was not just a national hero who fought against the enemy, but he was also the practitioner of the CCP's advocacy of army-people solidarity. The soldiers depicted in the Communist mass culture in Yan'an valued having an emotional bond with the peasants.

This chapter will first examine the CCP's wartime cultural and social policies in Yan'an to reveal its goals of managing close social integration and winning support from the majority of the society, the peasants. It aims to embed the discussion of the CCP's construction of the soldier figure in Yan'an into the political, social, and cultural background of wartime Yan'an. The second part of this chapter will provide a close reading of a representative genre of mass culture in wartime Yan'an, the *yangge* drama, to explore the depiction of the soldier figure in the mass culture.

The construction of the soldier figure in the *yangge* drama corresponded with a series of social campaigns launched by the CCP to reconstruct the social dynamics in its Yan'an base. The CCP's 1943 campaign of supporting the army was intended to be commensurate with the 1939 Great Production Campaign and the 1942–1944 Literary Rectification in order to unify the thoughts and the emotions of the masses and mobilize wider social participation. In the *yangge* dramas, cultural propagandists celebrated the emotional ties between soldiers and peasants as a defining factor in the soldier figure. The construction of the CCP soldier figure within the framework of army-people solidarity in wartime mass culture was crucial to the CCP's project of managing social integration in their revolutionary base.

Institutionalizing the Army-People Bond in Political and Social Movements

The CCP advocated its policies on army-people relations to support political, social, and economic campaigns that it launched. It was in these campaigns that the bond between soldiers and the people was institutionalized. Since the early years of the CCP army in 1927, the regulation of army-people relations had been an arena where the CCP practiced political education among soldiers in order to consolidate its rule among peasants. As early as the Autumn Harvest Uprising (*Qiushou qiyi*) that took place in the southern Hunan and Jiangxi Provinces in 1927, the first armed uprising led by the Communists, Mao Zedong stressed that the CCP's Red Army must not be isolated from the people and likened the army-people relationship to fish swimming in the sea. In the CCP's view, building the army did not just involve military training of soldiers' bodies; it also demanded regulating the military-civilian relationship by celebrating the close connection between soldiers and the masses.

The strategy that the CCP employed to build the soldiers' emotional bond with the peasants was to supervise strictly the everyday practices of the soldiers. During the period of the Jiangxi Soviet (1931–1934), the largest component territory of the Chinese Soviet Republic, the commander of the CCP army, Zhu De (1886–1976), developed several innovations for which the Red Army became known, one of which was the "Three Rules of Discipline and Eight Points of Attention" (*sanda jilü baxiang zhuyi*):

Three Rules
1. Obedience to orders.
2. Take not even a needle or thread from the people.
3. Turn in all confiscated goods.

Eight Points
1. Replace all doors and return all straw on which you sleep before leaving a house.
2. Speak courteously to the people and help them whenever possible.
3. Return all borrowed articles.
4. Pay for everything damaged.
5. Be honest in business transactions.
6. Be sanitary—dig latrines a safe distance from homes and fill them up with earth before leaving.
7. Never molest women.
8. Do not mistreat prisoners.[3]

Evans Fordyce Carlson (1896–1947), the famed US Marine Corps leader, documented the rules and regulations of the CCP army when he visited Yan'an in 1938.[4] Most of these rules in both the Jiangxi Soviet period and in the Second Sino-Japanese War period dealt with soldiers' interactions with civilians. The rules and regulations on soldiers' daily behavior attempted to cultivate a harmonious bond between the soldiers and the peasants.

In regulating military-civilian relations, the CCP was not only interested in disciplining the soldiers' behavior. What the CCP aimed to achieve in the base areas was wider social mobilization. To this end, the CCP tried to build two-way emotional communication between the soldiers and the peasants: soldiers should respect and love the people, and the people should support the soldiers.

To enlist participation from the masses in building army-people solidarity, the CCP-issued edicts on army benefits stipulated not only the favorable treatment that CCP soldiers could receive from the government but also the support they should get from the peasants. In December 1937, the Shaan-Gan-Ning Border Region government issued "Shaan-Gan-Ning bianqu kangri junren youdai tiaoli" (Edicts of treating favorably anti-Japanese soldiers in the Shaan-Gan-Ning Border Region), stipulating that anti-Japanese soldiers and their families should enjoy multiple benefits. For example, they should be exempted from all kinds of taxes, live in public-owned houses without paying rent, enjoy a 1 percent discount when purchasing in public-owned stores, be able to send their children to school for free, and receive free medical treatment if injured from war. Moreover, if there was an inadequate labor force to cultivate the land, they could request help from other people who resided in the border region.[5] The policy of treating anti-Japanese soldiers and their families favorably was integrated into the "Shaan-Gan-Ning bianqu kangzhan shiqi shizheng gangling" (Government guidelines in the Shaan-Gan-Ning Border Region during the Second Sino-Japanese War), issued in 1939.[6]

However, by the early 1940s the support for the army had faded among a number of the CCP members and the rural masses in the Border Region, largely due to years of living in relative peace. With the poverty of the border region exacerbated by blockade and war, and with the end of the subsidies from Chiang Kai-shek's Nationalist government for the Eighth Route Army[7] and the border region in 1941, economic options available to resistance forces became more narrow. Because of financial pressures, the tension between the army and the people loomed large. Many peasants who struggled to survive refused to provide economic or food support to the army. This situation was exacerbated by the actions of some soldiers, who forcibly took provisions from peasants without payment. The border region government, confronted with the threat to its wide social support from the increasing army-people conflict, launched the Rectification Movement (*Zhengfeng yundong*) in 1942 to consolidate intraparty consensus. The party also called for wider social participation in the Great Production Campaign (*Dashengchan yundong*) in the same year, in an attempt to promote a full-scale mobilization of the public sector in a drive for economic self-sufficiency.

The Literary Rectification ran parallel to the general political rectification of all party members. The Yan'an Forum on Literatures and Arts (1942–1944) was a crucial event in Chinese revolutionary culture. Mao Zedong's talks at this forum guided the direction of subsequent party policies on literatures and arts and shaped the cultural aesthetics of the intellectuals. Mao's talks set the tone for the propaganda culture in Yan'an by advocating "mass-ification" (*dazhonghua*) of literatures and arts. Mao clearly pointed out that the audience of art and literature in the base areas was supposed to be workers, peasants, and soldiers (*gongnongbing*) and their cadres (*geming ganbu*).[8]

David Holm, in his study of *yangge* in Yan'an, suggests a redefinition of Mao's concept of *dazhonghua*. In his opinion, *dazhonghua* was not a problem of how to put new content—the rudiments of the new learning, or "modern science," or an "advanced world view"—into a form accessible to the popular masses; instead Mao turned this around and said it was the writers and artists themselves who should be "mass-ified."[9] A close reading of Mao's talks shows that *dazhonghua* was in effect intended to regulate the emotions of the cultural workers and social intellectuals, such as the writers and the artists. Mao gave his own definitions of *dazhonghua* in his talks: "What is *dazhonghua*? It is that the thoughts and emotions of writers and artists should be fused with those of the masses of workers, peasants, and soldiers" (*wenyi gongzuozhede sixiang ganqing he gongnongbing dazhongde sixiang ganqing dacheng yipian*).[10] In other words, *dazhonghua* meant a complete integration in thoughts and emotions of the writers and artists with the masses.

Mao used his own examples of emotional changes (*ganqing bianhua*) to explain his standpoint that the peasants, workers, and common soldiers should be treated with high respect:

> I began life as a student, and at school I then used to feel it undignified to do even a little manual labor in the presence of my fellow students. . . . At that time I felt that intellectuals were the only clean people in the world, while in comparison workers and peasants were dirty. I did not mind wearing the clothes of other intellectuals, but I would not put on clothes belonging to a worker or peasant. But after I became a revolutionary and lived with workers and peasants and with soldiers of the revolutionary army . . . , I fundamentally changed the bourgeois and petty-bourgeois emotions implanted in me in the bourgeois school. I came to feel that the workers and peasants were cleaner than the bourgeois and petty-bourgeois intellectuals.[11]

In Mao's view, the intellectuals' backwardness was rooted in their indifferent and prejudiced emotions toward the peasants. Mao urged that if the writers and artists who came from the intelligentsia wanted their works to be well received by the masses, they must change and remold their thoughts and emotions.[12] Mao viewed emotional changes as class transformation.[13] In Mao's cultural ideology, every worker of literature and art should be a propagandist winning the support of the peasants and workers in the revolutionary base. Mao's outlook on the emotional change of the writers and artists was stressed by other party authorities. Zhou Yang, the head of the Lu Xun College of Arts and Literatures, stated that "it is critically important for workers of literatures and arts to transform their emotions."[14] For the CCP authorities, forging closer social integration in its base required a reform of thought and emotion among the writers and artists.

At the same time as the Literary Rectification, the CCP also turned its attention from political and military concerns to economic matters in an all-out effort to strengthen the base areas. The Great Production Campaign, which was initiated in 1939 and had greatly enlarged its scale by 1943, aimed at "the creation of a self-sufficient and more prosperous economy organized on cooperative, mobilization and participatory principles."[15] The CCP's production movement sought to make organized labor exchanges, like cooperatives and mutual-aid teams, the fundamental agricultural units in the base areas. The Literary Rectification and the Great Production Campaign both aimed to win wider social support. If the rectification could be seen as a cultural movement for the CCP to articulate the significance of the emotional bond with the peasants, then the production campaign was an economic movement to give the intellectuals and the soldiers a practical way to build their bond with the peasants. The rectification was designed to make the writers and artists understand, sympathize with, and respect the emotions of the peasants, while the production campaign was waged to have cultural workers and soldiers produce economic materials together with the peasants and thereby cultivate their emotional bond with them.

In the Jiangxi Soviet period, the army had played no part in agricultural production "because grain had been plentiful."[16] In Yan'an, the first region-wide production campaign was launched in 1939 to counter the Guomindang blockade, but it was only in the aftermath of the economic crisis and after the launching of the Rectification Movement that organized production developed throughout the border region.[17] The leadership particularly stressed the significance of the labor reserve in the army.[18] The engagement of the military in productive enterprises was not aimed only at improving the livelihood of the troops; the emotional attachment between soldiers and peasants was integrated into their slogans for their participation in production. The CCP army mobilized the soldiers to participate in the agricultural production campaign by evoking their sympathy toward the peasants. The army aimed at "a realization that the burden of the military on the strained financial resources of the people could be correspondingly lightened."[19] This rhetoric of lightening the burden of the military on the people demanded that soldiers respect and be considerate to the peasants; the soldiers built their bond with the people by participating in economic production and lowering the burden on them.

The two important movements in Yan'an—the Literary Rectification and the Great Production Campaign—were both justified by the authentic principle of the emotional bond between the peasants and other social forces, including the soldiers and the intellectuals. The CCP soldiers were expected to build solidarity with the peasants by regarding them with respect and understanding and by physically pursuing production work with them. The celebration of army-people solidarity was finally institutionalized in the 1943 Double-Supporting Movement (*shuangyong yundong*), which promoted a spirit of "supporting the army" (*yongbing*), "favorable treatment of soldiers' families" (*yongshu*), and the "'cherishing' of the people" (*aimin*).

In January 1943, Lin Boqu (1886–1960) and Li Dingming (1881–1947), respectively chairman and vice chairman of the border region government, issued "Yonghu jundui jueding" (Decisions on supporting the army) and "Bianqu zhengfu guanyu yongjun yundong yue de zhishi" (Directions of the border region government on the army supporting movement month). The CCP leaders also reformulated "Shaan-Gan-Ning bianqu kangshu lihun chuli banfa" (Methods of dealing with the divorce of anti-Japanese soldiers' wives) and "Youdai kangri junren jiashu tiaoli" (Edicts on treating favorably the families of anti-Japanese soldiers) as steps for further intervention into soldiers' marriages and family lives.[20] In the same year, the headquarters and the political department of the Rear Corps of the Eighth Route Army (*Balujun liushou bingtuan*) issued "Guanyu yonghu zhengfu aihu renmin de jueding" (Decisions on supporting the government and loving the people), the first pact on supporting the government and loving people in the history of the Chinese People's Liberation Army (PLA).[21] The issuing of these edicts marked the beginning of the Double-Supporting Movement. Later that year the movement spread to other CCP liberation areas, eventually becoming a tradition institutionalized in the history of the CCP to the present day.

The Double-Supporting Movement enlisted wide participation from the soldiers and rural masses in the CCP's wartime revolutionary bases. Substantial literary and artistic works of various genres were created with the theme of army-people solidarity. These cultural productions were published in various kinds of media and performed on diverse occasions; they were even written on village walls and read aloud by propaganda workers to illiterate people in village markets. These cultural works reached wide social audiences—most being common soldiers and local peasants—shaping their emotions and lives in many ways. As Holm argues, "a salient feature of the cultural workstyle that emerged after 1942 was the party's sponsorship of model genres that could be used as the basis for mass movements in culture."[22] *Yangge* is a representative of these model genres that eventually developed into a mass movement after 1943 because of its wide popularity in Yan'an. This mass movement involved extensive social participation mainly from the cultural workers and also from literate soldiers and peasants.

Articulating the Army-People Bond through the *Yangge* Movement

The previous section has shown that the CCP's Great Production Campaign, Literary Rectification, and Double-Supporting Movement in wartime Yan'an justify categorizing the army-peasant bond as a legitimate social relationship. Army-people solidarity became one of the dominant themes in cultural and propaganda works. This section focuses on how the CCP's cultural workers articulated the army-peasant bond in the *yangge* movement and how this articulation shaped the construction of the CCP soldier figure. To achieve this goal, this section will provide a close reading of some representative *yangge* plays.

The wide popularity of the *yangge* dramas among the soldiers can be partly explained by the low literacy rates and class backgrounds of the soldiers. Before 1931, the level of general education in the border region was very low; the literacy rate was only 1 percent.[23] When the Second Sino-Japanese War broke out in 1937, the majority of CCP soldiers were able to read about fifty words.[24] By 1941, the CCP army's anti-illiteracy efforts had developed to the extent that 80 percent of soldiers were able to recognize two hundred words.[25] In terms of class structure, 60 percent of the CCP soldiers were drawn from the peasants, 20 percent from the workers, and 20 percent from other groups, such as merchants and students. Ninety-five percent of the soldiers were young men aged between eighteen and thirty-five. Although the literacy rate in the CCP army had greatly improved in comparison with the prewar situation, the majority of soldiers were still unable to read newspapers, fiction, or other written forms of literature. The *yangge* drama, as an oral medium, became a preferable cultural genre among young soldiers.[26] The army often held night parties where the soldiers sang together and performed plays.[27]

Yangge is a form of traditional Chinese folk drama popular in northern China. Translated, the term means "the song sung while transplanting rice." According to incomplete statistics taken from CCP archives, the peak in performances of *yangge* on the theme of army-people solidarity was during the Double-Supporting Movement, which was first initiated during the 1943 Spring Festival.[28] The Northwest Bureau (*Xibeiju*) of the CCP called for a bigger-than-ever *yangge* movement in Yan'an, encouraging the formation of *yangge* propaganda troupes in every office and unit within the Yan'an area. As Holm argues, "the *Yangge* Movement, as it came to be called, was clearly given high priority by the Party leadership, and detailed pieces of reportage on the activities of individual *yangge* troupes, some by well-known literary figures, were published."[29] The *yangge* play writer and performer Wang Dahua (1919–1946) recalled that he often performed *yangge* in the street or in the square and that every performance had about ten thousand in the audience.[30] People in Yan'an were proud of the success of *yangge*, and whenever they were asked about literature and art in Yan'an, they would talk about *yangge*.[31]

A number of *yangge* troupes were organized by students in the Lu Xun College of Arts and Literatures (usually referred to as "Luyi") and other colleges in Yan'an, cultural workers in government agencies, army units, factories, and other social institutions, as well as literate peasants and soldiers. According to the CCP's own statistics, by 1944 there were 949 *yangge* troupes, which meant that, on average, there was a troupe for every fifteen hundred people.[32] Their performances attracted a cumulative audience of 8 million viewers.[33] Holm reveals that "open public performances of *yangge* by various troupes, whether together or separately, indoors or outdoors, were not uncommon around Yan'an and were probably the way in which the majority of the population saw the plays."[34] In 1943 more than 150 *yangge* dramas were performed. Among them, fifty-six described the theme of agricultural production, seventeen the theme of army-people solidarity, and ten the theme of self-defense.[35] It is hard to assess the completeness of these statistics, but it is reasonable to argue that production and army-people solidarity were among the two most popular themes in the *yangge* movement. The *yangge* dramas were not written by any individual; they were products of collective creation mainly by literature and art workers in Yan'an, who assumed the role of cultural propagandists.

The emotional ties between the soldiers and the peasants were highlighted in the *yangge* plays.[36] The 1942–1944 Literary Rectification was a watershed in the discourse on the soldier-peasant bond. The *yangge* dramas before the rectification movement tended to portray Communist soldiers as a force that morally transcended the peasants, physically protected the peasants, and offered the peasants more help than the peasants offered the soldiers. The *yangge* plays were full of sentences such as: "Every river has its source and every tree has its roots" (*shui you yuan, shu you gen*), or "the Eighth Route Army was the lifeblood of us peasants and our parents" (*balujun shi laobaixingde dajiuxing he qindieniang*).[37] Before the Literary Rectification, the *yangge* dramas that focused on the soldier-peasant relationship as the main theme highlighted the merits of the army by writing about the negative sides of the peasants. Such examples included *yangge* dramas complaining about how the peasants did not open up wasteland as fast as the soldiers or how the peasants were worried whether the soldiers would eat their food.

During the Literary Rectification, the tendency to highlight the army as a great force that guided and protected the people and to depict the peasants as backward and weak was criticized by CCP cultural leaders. Zhou Yang (1908–1989), for example, claimed that to describe the peasants in this way was to make fools of them.[38] Zhou pointed out that emphasizing the CCP army's role of protecting the peasants could not result in the desired army-people relationship. According to Zhou, this relationship should have two layers of meanings. Above all, the people were the roots, lifeblood, and parents of the army. By this understanding, the ideal relationship between soldiers and peasants was one similar to that between a son and a parent. The soldiers should regard the peasants as their parents, with high respect. The soldiers received not only physical care but also education on agricultural production work from the peasants. The celebration of the peasants' as soldiers' parents represented the CCP's efforts to pursue the continued popular support of male peasants.

One example of the *yangge* dramas dealing with the theme of the soldier-peasant relationship performed before the Literary Rectification was *Shi'er ba liandao* (Twelve sickles),[39] which was created in 1938. In this play, the political commissar (*zhengzhi weiyuan*) asks the peasant Wang Er to help make twelve sickles for the army so that the soldiers can participate in economic production. Wang Er's wife, who comes from outside the border region, strongly holds the traditional attitude toward the army and views soldiers as nothing but villains who bully the peasants. Wang Er is more than willing to accept the task given by the political commissar; he also educates his wife by explaining to her that the Eighth Route Army pursues the interests of the peasants and has deep love toward them.

The wife agrees with Wang's decision and gives him a hand by doing assistant work. However, she thinks that since the political commissar is very eager to have the sickles in a short period of time, they should charge him more money than the sickles are worth. She also tells Wang not to work too hard on a task given by the army official. Wang tells his wife that the army loves the peasants and thus helping army soldiers is something that the peasants should do. In this *yangge* play, the political commissar from the army only appears once, but the task he gives Wang Er is treated as an honorable mission by the peasant. Wang's wife, who has come from outside the border region, has "backward thoughts" (*luohou sixiang*) concerning soldier-peasant relations and thus needs to be educated.

The male peasant assumes the role of a propagandist and educator. His logic is that because the soldiers of the CCP army love the people and help the people fight against the Japanese, the people should help the army unconditionally. This *yangge* drama attempted to imbue into the peasants the sense that army soldiers were protectors of the people. The army's agenda of protecting the safety of the peasants was treated as the single embodiment of the soldiers' love for the people. The bond between soldiers and peasants was also gendered in this play; the army-people emotional ties were exclusively expressed between the soldier and the male peasant, yet were not embraced by the peasant wife.

After the Double-Supporting Movement was officially launched in the spring of 1943, the *yangge* movement reached its peak and even more *yangge* plays with the theme of the army-people bond were created. Among them, *Zhang Zhiguo* (The soldier Zhang Zhiguo)[40] and *Jun aimin, min yongjun* (The army loves the people, and the people support the army),[41] both performed in 1943, were the most popular. The soldier Zhang Zhiguo strongly resolves to be a model soldier by becoming a labor hero. Although his hands become swollen due to overwork at weeding, he still asks to join the army and weed every day. His director (*zhidaoyuan*) orders him to rest, but he refuses because he has resolved to surpass the production task so that the economic burden on the people can be lessened.

Another character in this play, the old peasant Tian, who is experienced in weeding, can remove thirty kilograms of weeds every day. When Tian hears that Zhang Zhiguo can remove 108 kilograms, he does not believe it and decides to go to the field and learn from Zhang in person. However, the soldier Zhang is very humble and asks Tian to become his teacher; Zhang also shows Tian his swollen hands. The old peasant feels extremely moved because he has not seen any army other than the Eighth Route Army that does production work to lessen the burden of the people. Tian is not only touched by the soldiers' enthusiasm of working on economic production but is also moved by the soldiers' humble and considerate attitude toward the peasants. This play shows that the model soldier ideal demanded maintaining an emotional bond with the people by actively participating in production work with the peasants and by adopting a humble and thoughtful attitude toward them.

The sentence "The army and the people had the same mind and were one family" (*junmin tongxin shi yijiaren*) appears repeatedly in this *yangge* play. "The same mind" shared by the soldiers and the peasants is not fighting but working together. The strong wish to do economic work justifies the soldier's intention of not following his leader's order. Zhang is not punished by his superior; instead he is praised and presented as a model for other soldiers to follow. Sharing experiences in agricultural production becomes the arena where soldiers and peasants build and strengthen their emotional ties, and participating in economic work becomes an indispensable factor in defining a soldier figure in this play.

In the *yangge* play *Jun aimin, min yongjun*, the wife of the peasant Wang considers the Eighth Route Army soldiers to be her saviors and wants to send dumplings and cloth to them as gifts. In her eyes, the Eighth Route Army soldiers are not only competent in fighting but also capable of doing production work. Her husband, Wang Er, also wants to be a model of supporting the army (*yongjun mofan*). The squad leader (*ban-zhang*) Wang comes to the peasant couple's place to drop off some firewood and invite them to attend the meeting promoting the campaign of supporting the government and cherishing the people (*yongzheng aimin hui*). Wang Er wants to give the gifts to the squad leader Wang, but the squad leader leaves quickly, pretending that he has a stomachache. In doing so, the squad leader Wang forgets his bag in the couple's house, so the peasant Wang puts the gifts inside the bag and gives them to him when he comes back to get his bag. However, when the squad leader Wang finds the goods in his bag, he secretly puts them in a basin at the door.

The peasant couple in this *yangge* play express their sincere affection toward the soldier by insisting on giving gifts to him; the squad leader also shows genuine emotion toward the peasants by refusing to accept the gifts without making the peasants offended. The emotional communication between the soldier and the peasants happens in the peasants' home. The way that they get along with each other by giving and refusing to accept gifts is like normal interactions among family members. Although the peasant couple in this play are both enthusiastic in supporting the army, the expression of emotions between soldiers and peasants is still gendered. The way the peasant wife shows her love for the soldier is still to cook for the family, although the definition of family is enlarged to include strange soldiers. The celebration of the army-people bond reinforced patriarchal values by stressing women's roles in performing housework.

During the high tide of the *yangge* movement in 1944, two famous *yangge* plays were created, *Liu Shunqing* (Soldier Liu Shunqing)[42] and *Niu Yonggui guacai* (The injured soldier Niu Yonggui).[43] In *Liu Shunqing*, two old peasants, Zhang and Li, admire the abilities of nineteen-year-old company commander Liu and his resolution in clearing wasteland and doing productive work. To support the army's agricultural production, they both give part of their own land to the army. Due to the lack of farming tools, Liu goes to consult with the two peasants. When Liu is worried that they cannot find metal to produce tools, the two peasants tell Liu that there are three old bells made of metal in the deserted village temple, but they do not believe that the three heavy bells can be made into tools of production. Liu believes that he can manage it as long as he can find a blacksmith. Liu first requests the local government to persuade the peasants to allow the soldiers to use the bells. He finds a blacksmith and asks him to go to work at the army base. Originally the blacksmith is not willing to go out to work, but the company commander keeps propagandizing the blacksmith, saying that the reason that the army is also doing farm production work is to help the peasants get rich. The blacksmith agrees with Liu, but is worried that the tools might not be finished within the requested seven days. Liu mobilizes his soldiers to assist the blacksmith. After the production tools are completed, the soldiers go to plow the farmland. Because of a lack of farm cattle, the soldiers do the plowing work that is normally performed by the cattle. The company commander Liu does not feel embarrassed when the two old peasants Zhang and Li see this situation. Instead he asks the two peasants to check whether the land is plowed well enough. The two peasants both think that the commander is the real hero, while Liu believes that the army's production totally depends on the help of the peasants.

The propaganda message that this *yangge* play conveyed was not simply that the army was the savior of the peasants and the peasants were grateful to the army. On the one hand, the soldier in this play, the company commander, has as much faith and passion in agricultural production as he does in fighting. On the other hand, the army's economic production and the making of its tools depend on the assistance of the masses. When the army lacks farming tools, they consult with the peasants. When they do not know how to make old bells into tools of production, they rely on the blacksmith. When they plow the farmland without using cattle, they ask the peasants to check on the quality of their work.

The two peasants, Zhang and Li, admire the company commander's passion in production, his creativity in resolving the lack of tools, and his spirit in plowing out the land without relying on cattle. Therefore, the two peasants view Liu as the real hero. However, Liu always relies on the peasants for help whenever any difficulty arises. The emotional bond of the soldiers with the peasants is embodied in Liu's humble attitude toward peasants and his readiness to learn from them. The situation in which the soldiers plowed the farmland without using cattle was documented by Nationalist and foreign journalists who visited Yan'an in 1944.[44] According to their observations, seeing the soldiers doing work that was normally done by animals led the peasants to overcome their hostility toward the soldiers.[45] Working on farm production not only became an indispensable quality of a soldier hero but also served as the best channel to strengthen the emotional ties with peasants. The commander's expression of respect for the peasants' authority in productive knowledge fulfills his image as a CCP soldier.

Niu Yonggui guacai is a 1944 *yangge* play of a common soldier called Niu Yonggui whose leg is injured in a battle. The peasant Zhao Shouyi has two sons: the elder one is beaten to death by the Japanese, while the younger one is sent to join the Eighth Route Army. Zhao Shouyi is grateful to the Eighth Route Army because the army killed his enemy. The injured soldier Niu Yonggui happens to arrive at Zhao's house, and Zhao insists that Niu should stay at his home, saying that he will protect Niu if the enemies come to search for him. The next morning, Niu Yonggui plans to leave because he is worried that he might get Zhao's family into trouble. However, upon Zhao's repeated requests, Niu goes down into a tunnel at Zhao's house. The Japanese beat Zhao harshly, asking him where the Eighth Route Army soldier is, but Zhao insists that he does not know. The Japanese finally leave. When Zhao is about to call Niu out of the tunnel, the Japanese, who know that Zhao has a tunnel at his home, return. The Japanese dare not enter the tunnel, so they force Zhao to shout toward the tunnel that the Japanese have left. No one comes out, so the Japanese leave again. After making sure the Japanese will not return, Zhao stamps his foot on the ground three times, and Niu finally emerges. Stamping three times was the signal Zhao had earlier told Niu would mean it was safe to come out. Under Zhao's help, Niu succeeds in returning to his troops.

In this play, the CCP soldier is an embodiment of the sincere army-peasant emotional bond. The soldier Niu Yonggui shows his care for the peasant; he is worried that he might get the peasant's family into trouble. Niu also receives the genuine love from the peasant Zhao, who is willing to risk his life to protect Niu. The soldier-peasant bond is justified by traditional father-son bonds in this play. Niu says that the Zhaos are his parents who gave him a second life (*zaisheng dieniang*), and Zhao replies that the Eighth Route Army soldiers are the children of the masses and that the masses should protect them.

A comparison of *Niu Yonggui guacai* and *Jun aimin, min yongjun* shows that the soldier-peasant intimacy is gendered in both plays. In the campaign of supporting the army, the peasant wife in *Jun aimin, min yongjun* can only play the leading role in housework by making dumplings and preparing other gifts for the soldiers. Women can only provide a supporting role in the fight against the Japanese. In *Niu Yonggui guacai,* it is the male peasant Zhao who risks his own life in the face of the Japanese to safeguard the life of the soldier. Zhao's wife is portrayed as a loyal yet cowardly woman who does not know how to handle the Japanese and protect the soldier. The army-people bond, which became an essential element in defining the soldier figure, reinforced women's supportive roles as caregivers. Furthermore, it reinforced the notion that the women were to be protected by their male counterparts. In this sense, the emotional bond between the soldiers and the peasants that was celebrated in mass culture strengthened the patriarchal values.

As Mao and his associates pressed for a new mass culture that drew on *yangge*, a new discourse on the soldier figure was forged along the lines of the army-people bond. The celebration of the cult of army-people solidarity in the CCP's significant social and political programs determined how the image of the soldier was shaped in mass culture. In these *yangge* plays that portray the soldier figure, both the camaraderie between soldiers and their feelings toward their lovers are obscured. The soldiers are all single males in these plays, and their social emotion has only one focus—attachment with the peasants. The examples of the *yangge* plays examined in this chapter that were performed after the Literary Rectification did not depict the soldier as a symbol of military strength in fighting; the standard for a real CCP soldier was that he possessed a deep sentiment with the peasants. In the *yangge* dramas, the soldiers were praised by the peasants as heroes not only because they fought against the Japanese but also because they humbly learned from the peasants, eagerly consulted with them, and worked hard on the farmland alongside them.

By pursuing production work with the peasants, the soldiers were building social connection and the affirmation of relationship in the most fundamental terms the peasants knew. As the anthropologist Sulamith Heins Potter concludes in her research on cultural construction of emotion in rural Chinese social life, in Chinese terms, "no mere emotional state or response could provide the wealth of meaning that measurable labor provides."[46] A Communist soldier called Zhou Xinfu confessed the dual nature of his occupation: he was both a peasant and a soldier.[47] The CCP soldier was expected to strengthen the emotional ties with the peasants by devoting himself to both fighting and producing. To do the peasants' occupation served as a catalyst in strengthening the emotional bond between the soldiers and the peasants.

The examination of the construction of the soldier figure in Yan'an's mass culture during the Second Sino-Japanese War shows that the CCP tried to discipline the emotions of the masses in achieving its state-building goals of building the army and forging closer social integration. It tried to make the cultural workers forge an emotional bond with the peasants, and make the soldiers and the peasants develop an emotional bond toward each other. Regulating the emotions of different social forces and transforming social relations were as crucial to the CCP's rise as the mobilization of a labor and fighting force. These efforts also helped the CCP enlist support from the peasants, the most numerous and most steadfast of the CCP's allies according to Mao.

During the Great Production Campaign, the intellectuals and the soldiers were mobilized to build an emotional bond by working together with the peasants; during the Literary Rectification, the intellectuals were commanded by the CCP to reform their emotions toward the masses of the people; during the Double-Supporting Movement, the soldiers and the peasants were encouraged to show their emotional support to each other. By launching these cultural and social movements, the CCP established the army-people bond as a legitimate category of social emotion in Yan'an. Intellectuals, who largely became the CCP's propaganda workers after the Literary Rectification, celebrated the army-people bond in the *yangge* movement—the bond that the CCP tried to build by launching the Great Production Campaign and the Double-Supporting Movement. The army-people bond was institutionalized by the CCP's campaigns into a legitimate category of social relations. The cultural workers, who reformed their emotions toward the peasants after the Literary Rectification, assumed the role of the cultural army of the CCP by propagandizing this bond in the *yangge* mass movement.

A comparison of the discourses on the soldier figure constructed by the GMD and the CCP shows that these two parties both valued the significance of military virtues for their state-building project of creating a strong army. The GMD defined the virtues of a model soldier as a code of ethics that was marked by his obedience to the political doctrine of the Three Principles of the People, his submission to discipline and regulation over personal behavior and emotional expression, and his subordination to Chiang Kai-shek as the leader of the hierarchical system. Chiang and the GMD state justified the virtue of political discipline and ideological indoctrination by appropriating the traditional principle of five components in warfare proposed in *Sunzi bingfa* (Master Sun's *Art of War*) and by trying to incite the knight-errant spirit among the nation's soldiers.

For the Communists in wartime Yan'an, the virtue that fulfilled a model

CCP soldier was a strong emotional tie with the peasants. The CCP had stressed the significance of army-people bonding ever since its army was built in 1927. When the CCP initiated the Great Production Campaign in 1939, it promoted this virtue by mobilizing social members, including the soldiers, to work on agricultural production with the peasants. In launching the Literary Rectification Movement in 1942, the CCP justified this virtue by demanding the complete integration in thoughts and emotions of the intellectuals with the peasants. The CCP further legitimated this virtue during the Double-Supporting Movement in 1943, which was intended specifically to strengthen the army-people solidarity. By launching a series of social and cultural campaigns in its revolutionary base, the CCP institutionalized army-people solidarity as an ultimate military virtue in the army.

The Maoist discourse on the soldier heroes in mass culture engineered social emotions between soldiers and peasants as a method of state-building alternate to that of the GMD. Mass culture in the CCP revolutionary base in Yan'an described the soldier figure not only as a national fighter but also as a farmland producer, just like the peasants. In the discourse on the soldier figure constructed by the CCP's mass culture in Yan'an, the soldiers received care, assistance, guidance, and education from their parents—the peasants—and thus treated the peasants with high respect. The CCP constructed the soldier figure to meet its goal of winning the support of the peasants it ruled in the revolutionary base. The mobilization of the social bond between the soldiers and the peasants was crucial for the rapid development of the CCP army and the social integration of the CCP's revolutionary base during the war.

Conclusion

This book focuses on the construction of the soldier figure between 1924 and 1945 by various political, social, and cultural forces. The forces examined in this book primarily include Chiang Kai-shek's Nationalist government, the Whampoa Military Academy, urban intellectuals, professionals, literary writers, and students in the Nationalist areas, as well as Chinese Communists in the revolutionary base of Yan'an. The discourses these forces constructed represented their various and sometimes conflicting goals, agendas, and strategies in state-building. The conceptualization of the soldier figure served as the ideological base on which Chinese Nationalists and Communists justified their legitimacy and strengthened their rule, and where the aforementioned nongovernmental urban publics asserted their political influence and voiced their own respective social concerns and criticisms of the GMD government. The various ways that these political, social, and cultural forces constructed the soldier figure reveal the divergent trajectories of state-society relations during the tumultuous period that China experienced during the first half of the twentieth century.

The state-building project of the Nationalists first demanded the creation of a strong army to defeat regional warlords and to unify China under the GMD banner. For Chiang Kai-shek, the process of building his Nanjing Nationalist state started with the foundation of the Whampoa Military Academy in 1924. As chapter 1 shows, the primary goal of this academy was to establish a reliable, well-trained, and politically indoctrinated armed force to support the Nationalist Party's objective of unification and salvation. The Nationalists intended to cultivate the cadets at Whampoa into model soldiers and, by extension, a highly capable cadre who could constitute and lead the national army. The strategy employed by Whampoa was to carry out civic education that fully regulated each cadet's thoughts, behaviors, emotional expression, and socialization.

As the commandant of Whampoa, Chiang Kai-shek exerted his strong influence on civic education through constant lecturing and admonishment to the cadets. According to his lectures, a model soldier for the national army should be reformed according to a code of ethics: he should arm his mind with the doctrine of the Three Principles of the People; he should unconditionally subject his daily life to military discipline; he should cultivate his morality by developing a heroic outlook on physical sacrifice; he should be a ferocious warrior with bravery and toughness to endure cruel military training; he should repress his emotional expression and limit it only to his nation, family, parents, and army colleagues; and he should show fraternity to his fellow soldiers and loyalty to Chiang as the national leader. The soldierly ideals advocated in this code were consolidated by military rules such as *Lianzuofa,* Chiang's personal interaction with Whampoa cadets, ritual and holiday commemoration, and the establishment of the Whampoa Alumni Association. In the discourse constructed by Chiang at Whampoa, bravery, military discipline, physical sacrifice, and army hierarchy were surrounded by a heroic, glorious, and romantic aura.

For Chiang Kai-shek, the model soldier trained at Whampoa would shoulder the responsibility of being a part of the national army and take on a leadership role that involved absolute submission to his chain of command within the academy and the army. Training loyal Whampoa cadets served Chiang's state-building goals of strengthening his own authority and expanding the influence of his central army into the forces of regional warlords. To this end, Chiang opened temporary training classes or branch campuses affiliated with Whampoa and recruited army officers from the regional warlords' armies. The regional warlords and common soldiers resisted the soldier ideal created by Chiang. They viewed Whampoa cadets not as model soldiers who could help national armies raise the professional standard but as Chiang's political tools to reinforce his rule and as formidable rivals who would threaten their own interests. Although some Whampoa graduates constituted the influential Whampoa clique in the central army, many of them were often victims of discrimination in the various warlords' armies.

The resistance from regional warlords that Chiang met in extending the soldier ideal to national armies reveals that the GMD government and army were unsuccessful in achieving total control of regional forces. Tension also existed within Chiang's army and at Whampoa. Many Whampoa cadets challenged the rigidity of military discipline and army hierarchy by requesting to defend their hometowns and demonstrating against the unfair aspects of the academy's administration. The emotional connections among Whampoa cadets through diverse networks also encouraged them to confront the academy authorities to protect fellow members in their groups.

The building of Whampoa was not limited to the Nationalists' initial state-building effort to create a strong army to achieve national salvation; it also served Chiang's personal state-building goal of consolidating his authority within the national armies. After Chiang's Nanjing Nationalist government was founded in 1927, particularly as the threats from the Chinese Communists and the Japanese increased in the 1930s, military mobilization and social control became Chiang's priorities in reinforcing the legitimacy of the GMD state. As chapter 2 shows, one effort made in the state-building project of militarizing and disciplining society was the issuing and implementation of the compulsory conscription system in the mid-1930s.

At the same time that the GMD government was designing its compulsory conscription system, it reintroduced the traditional local control model of mutual surveillance—the *baojia* system—between 1932 and 1934. This system, which was designed to militarize grassroots organizations, reflected the GMD's state-building strategies of expanding state institutions and penetrating the state's influence into local society. The *baojia* organizations, which performed the function of registering and administering military training of the able-bodied, were intended to serve as the basis for compulsory military service.

The GMD government issued the first compulsory conscription law on June 17, 1933, to complement the preexisting mercenary system. By stipulating military service as a legal duty of all male citizens, the GMD government advocated the militarized and disciplined citizen ideal. Several months after the conscription law was implemented on March 1, 1936, the GMD government made some notable revisions. It introduced the concepts of exemption from military service, deferment of military service, and prohibition of military service. The GMD's effort to revise the conscription law continued throughout the Second Sino-Japanese War. The revised 1943 conscription law made compulsory conscription the only military recruiting system in the GMD state. It reinforced the notion that performing military service would fulfill citizen status and that the citizen-soldier was a model citizen and a national hero to be emulated and respected by society. During the war, the GMD government also issued a series of edicts regarding the benefits and duties of soldiers as national citizens. These edicts revealed the GMD's effort to penetrate its influence into soldiers' personal lives.

The compulsory conscription law and its revisions institutionalized the link between the soldier and the citizen. To achieve military mobilization and social control, the GMD government tried to create a citizen ideal that was not only military-ready but also morally cultivated. The soldier was constructed by the GMD propaganda as a model citizen and the epitome of morality. Literary education in the army was conducted as citizen education, which included cultivating moral virtues for the soldiers. From 1934, the New Life Movement propagandists tried to justify the applicability of moral virtues advocated in the movement and celebrate the soldiers as the exemplar practitioners of them. The GMD's military laws, army education, and political propaganda, together with Chiang's civic education lectures at Whampoa, elevated the soldier's status as a model citizen and a national hero who was militarized by the Nationalist army, politicized with nationalist ideology and the Three Principles of the People, disciplined by military laws and army hierarchy, and morally cultivated by virtues advocated in the New Life Movement.

The celebration of soldier heroics in the GMD's discourse served the GMD's state-building goals of military mobilization and social control. However, the soldiers did not fully benefit from the elevation of their status in rhetoric; society did not highly respect them, nor did it feel inclined to emulate them in performing military service. The soldiers were paid poorly and badly fed, and their treatment deteriorated as the threats from the Japanese increased. The observable reality of soldiers' low status impeded active participation by society in performing mandatory military service. The implementation of the conscription system at local levels did not follow the principles of equality and fairness as prescribed and met resistance from both local residents and *baojia* heads. This resistance revealed the great tension between the GMD government and rural society.

As chapter 3 shows, although the GMD's discourse of the soldier figure was resisted by regional warlords and rural society, it was echoed by some urban intellectuals and professionals who participated in the state-building process by supporting the army and serving the soldiers after the Second Sino-Japanese War broke out in 1937. The commitment of these social forces to national salvation forged a wartime alliance between them and the GMD state. Many of these intellectuals who participated in war service were affiliated with the GMD government, and they propagated nationalist resistance to the masses when they visited the front line. They participated in army support and soldier service in various ways, such as reporting war stories from the front line, comforting and treating the wounded soldiers at hospitals, writing books summarizing their soldier support experiences, and offering suggestions.

In the war reportages, the intellectuals highlighted the soldiers' bravery and expressed their deep respect for it. The soldiers' bravery not only showcased their fearlessness but also their contempt for the enemy. In this sense, the writers collaborated with Chiang's discourse of celebrating the soldiers' heroics. However, by depicting the soldiers as ordinary human beings who suffered both physical and psychological pains in war, they diverged from the GMD's heroic soldier ideal. The soldiers did not refrain from expressing their emotions either. Their bravery was not described by the intellectuals to encourage the social masses to emulate the soldier and become a military-ready, politicized, and morally cultivated citizen. Instead, it was depicted as a contrast with the poor treatment of soldiers by the government. This contrast served the intellectuals' goals of justifying the importance of larger social participation in national affairs and asserting their own political influence as social mobilizers. It also allowed them to explicitly voice criticism of the GMD government's maladministration in organizing timely relief for the soldiers. In their view, it was not just physical injuries but also poor treatment from the government that led many soldiers to weak heartedness, irritability, and pessimism.

Like the intellectuals, urban professionals also actively participated in army support. They identified themselves not merely as social mobilizers but as army educators, and they aimed to serve as the bridge between the military and civilians. The Soldier Service Branch of the Chinese YMCA, which was a volunteer organization, served soldiers in various aspects of their lives. To meet the work agenda and style of this organization, the social activist Liu Liangmo described the image of the soldier as an affectionate and lonely human being who needed emotional support from civilians and also valued sincere friendship with them. Liu highlighted the soldiers' moral qualities, such as being sincere and valuing friendship; in this sense, he collaborated with the GMD's discourse of celebrating the soldier as the epitome of morality. However, he also complicated the GMD's soldier ideal by stressing that soldiers were poorly educated, especially when it came to knowledge of the nation and nationalism and on cooperating with the public. He also described wounded soldiers as tired and scared of war. Liu argued for establishing the Club for Wounded Soldiers to implement the strategy of education through entertainment. In advocating this strategy, Li criticized the high-handed measures adopted by many government bureaucrats in treating wounded soldiers and asserted the authority of urban professionals as army educators.

The urban professionals who claimed their role in influencing military education also included vocational educators. They argued that army education should cover professional skills from which the soldiers' long-term career development could benefit. The educational professional Chen Junming did not oppose the GMD government's expectation for wounded soldiers to return to the battlefield after recovery, but he argued that it was not practical for them to do so. In this way, Chen supported providing vocational training for wounded soldiers. The social activist Duan Shengwu, who served at the Political Branch of the Logistics Department of the Military Affairs Committee, referred to wounded and disabled soldiers as honorable soldiers, and he urged the government to cooperate with urban professionals in providing education and training for them. The urban professionals' and social activists' proposals for vocational rehabilitation for honorable soldiers were institutionalized by the GMD through the Association of Vocational Coordination for the Honorable Veterans in 1940 at Chongqing.

The GMD's effort in promoting vocational rehabilitation for honorable soldiers was supported by the educational professional Yu Zhaoming. Yu wrote a book on this undertaking at the government's request in 1942. He argued for greater social mobilization in the vocational rehabilitation programs and advocated that the government support and cooperate with social forces in this undertaking. To justify his goals, Yu first praised the soldiers for possessing moral qualities that were essential for them to navigate smooth transitions into civilian professions. In this sense, Yu echoed the GMD government's rhetoric for celebrating the soldier as the epitome of morality. However, Yu complicated the GMD's rhetoric by stressing that soldiers were poorly educated and thus lacked the knowledge and skills required for career-building. Yu claimed that soldiers should expel any sense of superiority over civilians. To demonstrate their important role as army educators, vocational rehabilitation professionals also pointed out how the function of honorable soldiers as moral models was very ambiguous; if they did not receive vocational rehabilitation, they might become a source of social disorder.

The urban intellectuals and professionals who provided support and education for soldiers both echoed and complicated the GMD's discourse of the soldier figure; in doing so, they advocated for their own political influence as social mobilizers and army educators. The ways that these social forces constructed the soldier figure reveal that their alliance with the GMD government was ambiguous. As chapter 4 shows, this ambiguous alliance with the government also existed among the literary writers Xiao Jun, Qiu Dongping, and Xie Bingying, who created imagery of guerrilla soldiers, lower-level Nationalist army officers, and female Nationalist soldiers in their fictional writings, battlefield reportages, and autobiographies.

Xiao Jun was a soldier-turned-writer with both military education and army service experience. His 1935 fiction *Bayuede xiangcun,* which depicted a small group of guerrilla soldier heroes in a Manchuria village, represented the fighting spirit of the rural peasants in the face of social crisis and foreign invasion—the ideal he had embraced since childhood. In revealing the great suffering and brave struggles of the social masses in this fiction, Xiao assumed the role of social critic and pursued the ideas of national independence and liberation. Xiao recognized the necessity for the social masses to be militarized in order to survive social and national crises. However, he questioned the GMD's heroic soldier discourse in many ways. In his fiction, national heroes who bravely defended the nation were not highly politicized, disciplined, and morally cultivated soldiers trained by the GMD army. Instead, the guerrilla unit was led by the Chinese Communists, and the soldiers were common people from the lower rungs of society who fought in order to achieve a better and more peaceful personal life. Even though these grassroots heroes did not receive formal military training, they possessed the determination and drive to not only fight bravely in combat but also to rely on each other for survival. Additionally, these peasant heroes did not restrain themselves from expressing their natural sexual desires.

The GMD's heroic soldier ideal was also subverted by Qiu Dongping's battlefield reportages published in *Qiyue,* a journal motivated by a strong desire for independence. Qiu joined the Nationalist army to defend Shanghai in 1937. His direct army experience allowed him to adopt a personal and critical perspective in describing the experiences, emotions, and mindsets of Nationalist army soldiers. In his battlefield reportages Qiu praised the bravery of the Nationalist army soldiers fighting against the Japanese and expressed his respect for them. In this sense, Qiu collaborated with the GMD's heroic soldier discourse. However, Qiu also questioned the GMD's soldier ideal by revealing the brutality of the war and the soldiers' suffering and faintheartedness during the battle.

In Qiu's reportages, the physical sacrifice, bravery, military discipline, and army hierarchy advocated by Chiang Kai-shek and the GMD government lost their glorious and heroic aura. War was absurd in that its violence not only gave the soldier a sense of strength but also made the soldier aware of his vulnerability. The soldier was treated by civilians as an information source to satisfy their curiosity about war rather than as a model citizen to emulate; his mental pain and emotional scars were hardly understandable to civilians. Qiu revealed the ironic aspect in the Nationalist army that many competent and brave soldiers were not killed by the enemy but instead at the hand of their own commanders simply for questioning their orders or challenging the hierarchy out of professional responsibility.

In questioning the GMD's heroic soldier ideal, Xiao Jun and Qiu Dongping both expressed a deep political awareness by exploring social crises caused by foreign invasion and the enemy's brutality. They also searched for how to build a strong army and nation in their writings; in this sense, they participated in the process of state-building. For Xiao, it was the social grassroots—the peasants—who constituted the vanguard of Chinese national resistance and the hope of Chinese recovery. If well mobilized, they could form the basis of a strong army. For Qiu Dongping, professional army officers with a strong sense of responsibility were those who could shoulder the mission of national salvation. They consciously conquered the distraction of personal emotions, such as attraction to women and music, anxiety, fear, and self-doubt. They valued the bonding with their fellow soldiers, fought the war bravely, and followed military orders loyally. They also maintained a rational mind, intellectual independence, and critical thinking toward disciplines and orders.

As chapter 5 shows, another social force that constructed the soldier figure in the GMD areas was educated youths, especially current students. They actively joined the Youth Army when Chiang Kai-shek and his GMD government launched the Campaign to Mobilize Educated Youths to Join the Army in late 1944. This campaign showed that the GMD's state-building efforts of militarizing society strengthened during the war. By performing military service in person, educated youths supported the GMD's state-building agenda of creating a strong army.

This campaign was intended to meet internal and external challenges in the later years of the Second Sino-Japanese War. The GMD army was confronted with severe difficulties in recruiting new conscripts and decreasing combat capacity. As China's Second Sino-Japanese War was integrated into the greater World War II after the Pacific War broke out in 1941, Chiang Kai-shek became commander-in-chief of the Allied forces in China, Vietnam, Burma (present-day Myanmar), and Thailand. The increased collaboration with the Allied forces and the equipment and personnel support from abroad heightened his need for well-educated soldiers. China's resistance efforts met serious challenges after the loss of Operation Ichigō in 1944. All these factors led Chiang and his GMD government to strengthen their efforts in militarizing educated youths, specifically high school and university students—the social group that was exempted from performing military service by the 1933 compulsory conscription law.

The GMD reversed the exemption of students from performing military duty in March 1943 in the revised conscription law. In October 1944, the GMD held a conference that officially started the campaign to mobilize educated youths to join the army. To attract their participation, Chiang employed several forms of rhetoric to enhance the significance of the Youth Army. He first elevated the status of the Youth Army as the model and backbone of all national armies. The ability of the Youth Army soldiers to judge the enemy's situation independently promoted the fighting spirit and combat capacity of national army units. Chiang also stated that military knowledge and combat experience were indispensable to the youth's study and development. The Youth Army was advocated as the best professional school, where the youth learned practical skills and developed social leadership. Chiang argued that military training, which would advance the youth's own academic studies and career pursuits, was a significant part of national education.

Educated youths, particularly the high school and university students, actively joined the army upon the GMD's military mobilization efforts. They supported the GMD's army-building efforts and agreed with Chiang's rhetoric that justified the campaign for militarizing students. They claimed in their writings that their army experience was one way to enhance their knowledge and improve their physique. However, they argued for a soldier ideal that differed from the one created by Chiang and his GMD government; they portrayed themselves as the practitioners of the principles of self-government and democracy. In this manner, they applied these principles not only in their studies and entertainment experiences, but also in the role of army administration. Doing so encouraged the soldiers in the Youth Army to argue for an equal officer-soldier relationship even though they did not win any compromise from the army officers who directly led them. The pursuit of self-government and democracy also encouraged educated youths to voice their dissatisfaction with the political and military authorities. The educated youths who joined the Youth Army asserted a soldier ideal that was based on active and full participation in army administration.

Construction of the soldier figure was practiced not only by the GMD government and the intellectuals, professionals, writers, and students in GMD areas who sought to assert their state-building goals in regard to creating and supporting the army. As chapter 6 shows, during the Second Sino-Japanese War the CCP in the revolutionary base of Yan'an also created a discourse of the soldier figure in pursuit of its state-building goals of winning support from the largest population in the base—the peasants—and managing closer social integration.

In order to ensure continued popular support for military recruitment, the Chinese Communists made strenuous efforts to build army-civilian solidarity in Yan'an. To this end, the CCP actively united the peasants to justify its social, economic, and cultural movements and remold the emotions of the intellectuals and the CCP soldiers. To forge an emotional bond with the peasants, the CCP strictly disciplined the daily behavior of its soldiers related to their interactions with the peasants. The CCP not only demanded that the soldiers respect and love the peasants, but also encouraged the peasants to support the soldiers.

To further discipline the thoughts and emotions of all in Yan'an, the CCP carried out the Literary Rectification Movement and the Great Production Campaign in 1942. Mao's talks during the rectification set the tone for the cultural intellectuals' production and aimed to forge complete alignment in thoughts and emotions between the intellectuals and the peasants. The Great Production Campaign was intended to let the intellectuals and the soldiers consolidate their emotional ties with the peasants by performing production work together. The CCP's effort of forging emotional communication between soldiers and peasants reached its peak in the 1943 Double-Supporting Movement. These cultural, economic, and social movements launched by the CCP in wartime Yan'an were intended to validate the authentic principle of the emotional bond between the peasants and the soldiers and justify the army-peasant bond as a legitimate social relationship.

An examination of the *yangge* dramas shows that army-peasant solidarity was one of the dominant themes in the cultural works in wartime Yan'an. The different ways that the *yangge* plays depicted soldier-peasant relations before and after the 1942 rectification show that cultural workers remolded their own thoughts based on the emotional bond with the peasants. The *yangge* dramas before the rectification tended to highlight the army as a great force for guiding and protecting backward-thinking peasants. However, the *yangge* dramas created during and after the rectification stressed that the soldiers should highly respect the peasants and regard them as their parents. The soldiers received physical care and education on agricultural production from the peasants. The emotions between soldiers and peasants were thus engineered by the CCP's discourse on the soldier figure in mass culture to serve its state-building goals of winning the support from the peasants and managing closer social integration.

The construction of the soldier and social participation in state-building processes regarding how to build and support the army were highly gendered. The GMD state stipulated in its 1933 compulsory conscription law that military service was only a duty of male citizens; the revised 1943 conscription law claimed that women also had military duties, but theirs were limited to noncombat assignments. Although women intellectuals and activists like Xie Bingying did not challenge the GMD's stipulation, they asserted their political influence as social mobilizers by describing the soldiers as fragile and affectionate and in need of care from the larger society, even children. Xie, who was a woman soldier with Whampoa experience, did not depict the soldier as vulnerable to war brutality, as the male writers Xiao Jun and Qiu Dongping did. Instead, she portrayed herself as a brave rebel who opposed traditional gender roles and struggled for independence. In Xiao Jun's fictional *Bayuede xiangcun* (Village in August), war trauma was gendered, as women's suffering included sexual assault. Men fought the war to protect their lovers and for a better life for their families, while women joined the army unit because they had no choice and were overwhelmed with the resolution to revenge the killing of their families by the Japanese.

The construction of the soldier in the CCP's discourse was also gendered. In the *yangge* dramas celebrating the army-people bond, male peasants wholeheartedly supported soldiers and were brave enough to fight the Japanese to protect them. In contrast, women peasants in the *yangge* plays either were portrayed as being morally backward or supporting the soldiers only by performing housework. The gendered role of the social masses in supporting the army revealed the CCP's intention of winning male peasants.

The political culture surrounding the construction of the soldier figure shows that the GMD and CCP governments, intellectuals, professionals, writers, and students all had different state-building agendas and priorities. Treating state-building the same as the process of nation-building, Chinese Nationalists intended to build a strong and loyal army to unify China and consolidate the rule and legitimacy of the GMD regime, and to penetrate its influence into regional warlords' armies and rural society. Chiang Kai-shek also aimed to strengthen his own authority within the army and the society. Conversely, Chinese Communists aimed at building a state at the revolutionary base and considered their priorities to be winning the support of peasants, who constituted the majority of the population, and forging close social integration by reforming the thoughts and emotions of social members in the base.

Separate from the parties' apparatuses, the intellectuals, professionals, writers, and students discussed here embraced the nationalistic appeal and considered contributing to national salvation and asserting their political influence to be priorities of state-building. The intellectuals and professionals supported the Nationalist army by reporting the war from the front line, serving wounded soldiers at hospitals, and writing pamphlets on soldier support experiences and plans. They argued for their influence as social mobilizers and army educators when participating in the state-building processes. The literary writers revealed the brutality of the war, the cruelty of the enemy, and the suffering of the soldiers. They asserted their role as social critics and searched for the potential forces that would constitute a strong Nationalist army. The educated youths, specifically students, actively joined the Youth Army upon the GMD government's military mobilization, helping build the army to meet domestic and international challenges. They argued for the practice of self-government and democracy as soldier ideals in their published writings.

State-society relations engendered during the state-building processes in regard to creating and supporting a strong army presented different trajectories. The GMD's soldier discourse that advocated Whampoa cadets as model soldiers was resisted by regional warlords, who viewed the cadets trained at Whampoa as Chiang's personal tools to broaden his influence. The GMD's highly disciplined soldier ideal was also challenged by some Whampoa cadets, who tried to uphold their own interests. The GMD's rhetoric elevating soldiers' status to that of model citizens and the epitome of morality was also largely resisted by rural society organized in the *baojia* system, who viewed the soldiers' lives as miserable.

The GMD's heroic soldier ideal and rhetoric justifying the militarization of society was echoed by urban intellectuals, professionals, writers, and students. They praised the bravery and dutifulness of the Nationalist soldiers and recognized the necessity of militarizing the society, including rural peasants and educated youth. However, their echoing of the GMD's discourses of the soldier figure was not done with the intention of forming a strong alliance with the GMD government. Instead, they complicated, de-idealized, or confronted the GMD's soldier ideals to argue for their own political agendas and assert their own influence. While the intellectuals in the GMD areas constructed a soldier figure that did not fit the GMD's soldier ideals, the construction of the soldier in the mass culture in Yan'an served as propaganda supporting the CCP's army-peasant solidarity ideal. In this sense, the CCP managed closer social integration than the GMD during the Second Sino-Japanese War. In closing, this book has shown that the construction of the soldier figure was shaped by the various forms of participation in the state-building processes by multiple forces, revealing that there existed diverse trajectories in state-society relations in modern China.

Glossary

Balujun liushou bingtuan	(Rear Corps of the Eighth Route Army)
Beifa	(Northern Expedition)
Benxiao kaishi jinianri	(Commemoration of the Foundation of the [Whampoa Military] Academy, June 16)
Bianqianqu	(reorganized military zone)
Dashengchan yundong	(Great Production Campaign)
Dongzheng	(Eastern Expedition)
Erdengbing	(private E-1)
Guochi jinianri	(Commemoration of National Shame, May 9)
Guomin canzhenghui	(National Political Council Conference)
Guomin gemingjun	(National Revolutionary Army)
Guominbing	(citizen-soldiers)
Guonanxiang	(national-calamity-pay)
Huanghuagang lieshi jinianri	(Commemoration of the Huanghuagang Martyrs, March 29)
Huangpu junxiao	(Whampoa Military Academy)
Huangpu junxiao biyesheng diaochake	(Investigation Office of Whampoa Academy Graduates)
Huangpu junxiao jiaodaotuan	(Training Regiment of Whampoa Military Academy)
Huangpu tongxuehui	(Whampoa Alumni Association)
Huangpu xi	(Whampoa Clique)
Huanyi	(deferment of military service)
Jingshen jiaoyu	(civic education)
Jinyi	(prohibition of military service)

Kangri minzu tongyi zhanxian	(Anti-Japanese National United Front)
Kangri zhanzheng	(Anti-Japanese War or the Second Sino-Japanese War)
Laodongjie	(Labor Day, May 1)
Lian E rong Gong	(alliance with the Soviet Union and the CCP)
Lianzuofa	(law of joint responsibility)
Liao dangdaibiao beibu jinianri	(Commemoration of the Capture of Party Representative Liao Zhongkai, August 20)
Mianyi	(exemption from military service)
Qingnianjun	(Youth Army)
Quanguo gejie jiuguo lianhehui	(United Association of Various Circles for National Salvation)
Rongyu junren zhiye xiedaohui	(Association of Vocational Coordination for the Honorable Veterans)
Sanda jilü baxiang zhuyi	(Three Rules of Discipline and Eight Points of Attention)
Sanmin zhuyi	(Three Principles of the People)
Sanping yuanze	(Principle of the Three Equals)
Shaan-Gan-Ning *bianqu*	(Shaanxi-Gansu-Ningxia Border Region)
Shaji can'an jinianri	(Commemoration of Shaji Massacre, June 23)
Shangdengbing	(private first class E-3)
Shanghai can'an jinianri	(Commemoration of the Shanghai Massacre, May 30)
Shangshi	(staff sergeant)
Shuangyong yundong	(Double-Supporting Movement)
Sichuan tongxianghui zhandi fuwutuan	(War Service Corps of Sichuan Native Place Association)
Tongmenghui	(Revolutionary Alliance)
Wenyi zhengfeng	(Literary Rectification)
Xiangtu xiaoshuo	(countryside fiction)
Xiashi	(corporal)
Xinshenghuo yundong	(New Life Movement)
Yidengbing	(private E-2)
Yiwu zhengbingfa	(compulsory conscription law)
Youxia	(knight-errant)

Zhandi baogao wenxue	(battlefield reportage)
Zhandi fuwutuan	(War Service Corps)
Zhishi qingnian	(educated youths)
Zhishi qingnian congjun yundong	(Campaign to Mobilize Educated Youths to Join the Army)
Zhishi qingnian yuanzhengjun	(Educated Youth Expedition Army)
Zhongguo qingnian jidujiao xiehui junren fuwubu	(Emergency Service to Soldiers Program of the Chinese Young Men's Christian Association)
Zhonghua zhiye jiaoyushe	(Society for Chinese Vocational Education)
Zhongshi	(sergeant)
Zhongyang lujun junguan xuexiao	(Central Army Officer Academy)
Zongli danchen jinianri	(Anniversary of Premier Sun's Birthday, November 12)
Zongli jinian zhou	(Premier Sun Yat-sen weekly memorial service)
Zongli shishi jinianri	(Commemoration of Premier Sun Yat-sen's Death, March 12)
Zongli yixun	(Last Testament of Premier Sun Yat-sen)

Acknowledgments

The completion of this book involved a long journey of intellectual inquiry. When I grew up in China during the 1990s, I enjoyed watching war films on television. The older military movies produced in the early years of the People's Republic of China tended to construct the stereotype of the soldier as a paragon of strength and virtue. A new trend of war films appeared in China during the 1990s, when the films started to delve into the complicated theme of diverse individual experiences during the war, depicting the soldier as a multifaceted and complex figure. However, it was not until I attended The Ohio State University that my general interest in the imagery of Chinese soldiers became one of scholarly inquiry and academic research. I will always be indebted to the stimulating academic community of East Asian Studies at OSU.

I owe my greatest intellectual debt to Christopher A. Reed, whose scholarship and academic vision had a strong influence on the theme and structure of this book. It was his approach to China's modern media that inspired me to work on the topic of the extensive imagery of soldiers to shed new light on the state-building and state-society relations in modern China. He was very generous in sharing his vast knowledge on archival and database sources. Without his unfailing support, genuine care, encouragement, and faith in me, I would not have been able to accomplish this work.

My heartfelt gratitude also goes to Patricia Sieber and Ying Zhang for their generous help and encouragement. Professor Sieber's expertise on Chinese literature enlightened me on how to use literary works as primary sources to analyze historical issues. Professor Zhang provided detailed comments on my manuscript draft and helped me articulate the thesis succinctly. I would also like to thank other esteemed professors: Cynthia Brokaw, Philip Brown, James Bartholomew, and Judy Wu. Their preeminent scholarship and the stimulating discussions at their history seminars refreshed my understanding on a variety of topics and increased my critical and analytical abilities.

In the process of writing this book, I have received many invaluable comments from colleagues. Particular gratitude goes to Edward McCord, Xiaobing Li, Alan Baumler, Qiang Fang, Peter Harmsen, and Richard Frank for reading parts of my work. I am grateful to the CCKF-CHCI Summer Institute *China in a Global World War II* held at the University of Cambridge in July 2017. The conveners Hans van de Ven and Yeh Wen-hsin and other scholars listened to my manuscript presentation and provided many useful insights and suggestions.

Special thanks go to my colleagues at Spelman College. Kathleen Phillips-Lewis, who served as chair of the History Department when I first joined Spelman in 2014, Dalila de Sousa, my present department chair, Tinaz Pavri, chair of the Asian Studies Program, Cynthia Spence, UNCF/Mellon Programs Director, and Dimeji Togunde, Associate Provost for Global Education, helped me get several grants for my book project, including a UNCF/Mellon Domestic Faculty Residency Fellowship, a Department of Education Title III Grant, a Faculty Development Grant, and a Course Development Grant for Globalizing the Curriculum in Asian Studies. These grants allowed me to conduct research at the Fairbank Center for Chinese Studies at Harvard University, Stanford University, the University of Minnesota, and libraries and archives in Shanghai, Nanjing, and Taipei. Professor de Sousa helped me get a junior faculty research leave, which freed me from my teaching and service load in the fall semester of 2017 so that I could concentrate on the manuscript revision. She also skillfully guided me through the final stages of completing this work. My other History Department colleagues, including James Gillam, Margery Ganz, Charissa Threat, Nafeesa Muhammad, and Catherine Odari, as well as my Asian Studies Committee colleagues, offered their comments and suggestions on research-teaching balance, grant applications, and the publication process. I would also like to express my gratitude to my previous and present department secretaries, R. Nicole Smith, Antisha Burns, and Mia Mathis, who helped me with travel arrangements and reimbursement, grant applications, mail deliveries, photocopying, and class paperwork.

I greatly appreciate the comments and critiques I received from all of the above professors as well as the opinions and ideas shared by numerous colleagues and friends. Nonetheless, none of them is responsible for any errors or omissions in the book.

Many thanks are due to the staff of the following libraries and archives for their kind assistance in finding many precious documents that enriched my research: Harvard-Yenching Library, the Hoover Institution, the University of Minnesota Libraries, the Shanghai Municipal Library and Archives, the Second Historical Archives of China in Nanjing, and the Guomindang Party Archives in Taibei. Special appreciation also goes to Shulin Tan, my professor at Nanjing University who wrote a reference letter and helped me use the archives in China.

It was my great pleasure to work with the University Press of Kentucky editors, who provided gracious and professional assistance during the publication process. Senior editor Shiping Hua selected this manuscript and believed in its value. Allison Webster and Melissa Hammer, previous and present acquisitions editors, Natalie O'Neal, acquisitions assistant, Ila McEntire, senior editing supervisor, and Jackie Wilson, marketing assistant, all worked patiently on the manuscript. Derik Shelor spent several months editing my manuscript, offering me valuable suggestions for changes and rephrasing. I also would like to thank the anonymous readers selected by UPK. Their critical comments and constructive suggestions have made this a better book.

This acknowledgement would not be complete without directing my deepest gratitude to my family. I sincerely thank my parents, Zhongxin Xu and Baoling Liu, and my brother, Renyou Xu, for their boundless support and unconditional love. With their support, I have been able to release myself from many family duties and concentrate on my writing and career. My parents' spirit and their expectations for me have always been the inspiration for my life. This work is the best memorial that I could give them. I would like next to mention my husband, Sean Chao. His extensive knowledge of Chinese politics, history, and culture has opened the window for me to see how Americans have viewed issues related to China. His wit has kept me smiling, and his sarcasm has given me a different view of the world that has helped me keep things in perspective. Through his love, support, and unwavering belief in me, I have been able to complete this long journey. Through his eyes I have envisioned myself as a capable, intelligent woman who can do anything once I make up my mind.

I must also mention my three daughters, Evelyn, Elaine, and Elise. They have been the twinkle in my eyes since they were born. Whenever I feel tired and stressed, their big smiles and sweet words give me a much-needed reprieve from work and encourage me to pursue a better life for them. They are my angels. I love them very much. I hope that my perseverance in my work commitment will encourage them to be strong and inspire them to tackle any challenge in their life as they grow up.

Notes

Introduction

1. Andrew Huebner, *The Warrior Image: Soldiers in American Culture from the Second World War to the Vietnam Era* (Chapel Hill: Univ. of North Carolina Press, 2008), 2.

2. The Second Sino-Japanese War is also referred to by the Chinese as the Anti-Japanese War or the Chinese People's War of Resistance against Japanese Aggression.

3. Huebner, *The Warrior Image*, 1.

4. Most recent works of scholarship on twentieth-century China that merge military, social, and cultural history and tell stories of common soldiers include Diana Lary, *China's Civil War: A Social History, 1945–1949* (Cambridge, UK: Cambridge Univ. Press, 2015); Li Xiaobing, *A History of the Modern Chinese Army* (Lexington: Univ. Press of Kentucky, 2007); Neil Diamant, *Embattled Glory: Veterans, Military Families, and the Politics of Patriotism in China, 1949–2007* (Lanham, Md.: Rowman and Littlefield, 2009).

5. For the convenience of writing, the book refers to the Whampoa Military Academy as the Whampoa Academy, the academy, or simply Whampoa.

6. The term "urban publics" is drawn from Peter Zarrow's words that "[in Chiang Kai-shek's Nanjing regime,] both liberal intellectuals and nationalist students represented urban public opinion." See Peter Zarrow, *China in War and Revolution, 1895–1949* (New York: Routledge, 2005), 264.

7. In her study of warlord soldiers between 1911 and 1937, Diana Lary summarizes Chinese armed men into the following categories: major armies (multi-province warlord cliques, the GMD central army), local armies (single province or multi-county armies), petty armies (single county armies), militias (local defense forces at the county level, merchant-raised defense forces), bandit gangs (land-based bandit gangs and pirates), irregular units (temporary units raised from men normally under arms as bandits or secret society members), mass units (forces of

armed peasants, sometimes members of secret societies, sometimes political activists), stragglers (individual soldiers detached from their original units), and local bullies (village enforcers). See Diana Lary, *Warlord Soldiers: Chinese Common Soldiers, 1911–1937* (Cambridge, UK: Cambridge Univ. Press, 1985), 3.

8. The National Revolutionary Army was shortened to the Revolutionary Army (*Gemingjun*) before 1928 and the National Army (*Guojun*) between 1928 and 1949. During the Second Sino-Japanese War, the armed forces of the CCP were nominally incorporated into the National Army while retaining separate commands, but they broke away shortly after the end of the war. After Chinese Nationalists retreated to Taiwan in 1949, the National Army was renamed the Republic of China Army Forces (*Zhonghua minguo guojun*).

9. Zarrow, *China in War and Revolution,* 249. Zarrow also listed several examples of rebellions against Chiang Kai-shek's Nationalist government: Guangxi generals in 1929, Generals Feng Yuxiang (1882–1948) and Yan Xishan (1883–1960) in 1930, a growing Communist insurgency in Jiangxi in the early 1930s, Guangzhou supporters of the conservative GMD leader Hu Hanmin (1879–1936) in 1931, Fujian dissidents in 1933, and Guangdong and Guangxi forces again in 1936.

10. Ibid.

11. Ibid.

12. According to the Military Affairs Commission's 1928 organization law, "the chairman of the Military Affairs Commission, in shouldering his full responsibility of national defense, shall have supreme command of the land, naval and air forces, and shall direct the people of the entire nation." See Lloyd Eastman, "Nationalist China during the Sino-Japanese War, 1937–1945," in *The Nationalist Era in China, 1927–1949,* ed. Lloyd Eastman et al. (New York: Cambridge Univ. Press, 1991), 127.

13. Ibid., 26.

14. Leo Ou-fan Lee, "Literary Trends: The Road to Revolution," in *An Intellectual History of Modern China,* ed. Merle Goldman and Leo Ou-fan Lee (New York: Cambridge Univ. Press, 2002), 196.

15. Jing Wang, *When "I" Was Born: Women's Autobiography in Modern China* (Madison: Univ. of Wisconsin Press, 2008), 5. Another study of women's autobiography is Lingzhen Wang, *Personal Matters: Women's Autobiographical Practice in Twentieth-Century China* (Stanford, Calif.: Stanford Univ. Press, 2004).

16. Wang, *When "I" Was Born,* 189.

17. Lee, "Literary Trends: The Road to Revolution, 1927–1949," 229.

18. Ibid., 242.

19. Wen-hsin Yeh, "Writing in Wartime China: Chongqing, Shanghai, and Southern Zhejiang," paper presented at the CCKF (Chiang Ching-kuo Foundation for International Scholarly Exchange)-CHCI (Consortium of Humanities Centers and Institutes) Summer Institute, *China in a Global World War II,* Cambridge, UK, July 2017, 2.

20. Lee, "Literary Trends: The Road to Revolution, 1927–1949," 242.

21. Chih-tsing Hsia, *A History of Modern Chinese Fiction* (Bloomington: Indiana Univ. Press, 1999), 386.

22. Lee, "Literary Trends: The Road to Revolution, 1927–1949," 245.

23. Charles A. Laughlin understands reportage to mean "any deliberately literary nonfiction text that narrates or describes a current event, person, or social phenomenon." See Charles A. Laughlin, *Chinese Reportage: The Aesthetics of Historical Experience* (Durham, N.C.: Duke Univ. Press, 2002), 2. This book adopts Laughlin's definition of reportage. For studies on Chinese reportage literature, also see Yunzhong Shu, *Buglers on the Home Front: The Wartime Practice of the Qiyue School* (Albany: State Univ. of New York Press, 2000).

24. Lee, "Literary Trends: The Road to Revolution, 1927–1949," 202.

25. Ibid., 204. As Lee notes, the guideline did not define the scope of proletarian literature exclusively in the framework of workers, peasants, and soldiers as Mao later did in the 1942 Yan'an talks.

26. Ibid., 217.

27. The masterpiece on the CCP's Literary Rectification is Gao Hua, *Hong taiyang shi zenyang shengqi de: Yan'an zhengfeng yundong de lailong qumai* (How the red sun rose: The origins and development of the Yan'an Rectification Movement) (Hong Kong: Zhongwen daxue chubanshe, 2000).

28. Yinghong Cheng, *Creating the "New Man": From Enlightenment Ideals to Socialist Realities* (Honolulu: Univ. of Hawai'i Press, 2009), 59.

29. Lorenz Bichler, "Coming to Terms with a Term: Notes on the History of the Use of Socialist Realism in China," in *In the Party Spirit: Socialist Realism and Literary Practice in the Soviet Union, East Germany and China,* ed. Hilary Chung, Michael Falchikov, B. S. McDougall, and K. McPherson (Amsterdam: Rodopi B. V., 1996), 34.

30. Xiao Yanzhong, "Recent Mao Zedong Scholarship in China," in *A Critical Introduction to Mao,* ed. Timothy Creek (New York: Cambridge Univ. Press, 2000), 285.

31. Xiaoqun Xu, *Chinese Professionals and the Republican State: The Rise of Professional Associations in Shanghai, 1912–1937* (New York: Cambridge Univ. Press, 2004), 1.

32. Yu Zhaoming's birth and death years are unclear.

33. Suzanne Pepper, *Civil War in China: The Political Struggle, 1945–1949,* 2nd ed. (Lanham, Md.: Rowman and Littlefield, 1999), 426.

34. Ibid., 42.

35. Lloyd Eastman, *Seeds of Destruction: Nationalist China in War and Revolution, 1937–1949* (Stanford, Calif.: Stanford Univ. Press, 1984).

36. Julia Strauss, *Strong Institutions in Weak Polities: State Building in Republican China, 1927–1940* (Oxford: Clarendon Press, 1998), 2.

37. Ibid., 8.

38. Ibid., 184.

39. Ibid., 191.

40. Eastman, *Seeds of Destruction,* 219.

41. Strauss, *Strong Institutions in Weak Polities,* 56.

42. Arthur Young, *China's Nation Building Effort, 1927–1937: The Financial and Economic Papers* (Stanford, Calif.: Hoover Institution Press, 1971).

43. David Strand, "'A High Place Is No Better Than a Low Place': The City in the Making of Modern China," in *Becoming Chinese: Passages to Modernity and Beyond,* ed. Wen-hsin Yeh (Berkeley: Univ. of California Press, 2000), 99.

44. Morris Bian, "Building State Structure: Guomindang Institutional Rationalization during the Sino-Japanese War, 1937–1945," *Modern China* 31, no. 1 (January 2005): 37.

45. Ibid.

46. Morris Bian, *The Making of the State Enterprise System in Modern China: The Dynamics of Institutional Change* (Cambridge, Mass.: Harvard Univ. Press, 2005), 14.

47. The term "educated youth" (*zhishi qingnian*) mainly refers to high school and college students, young professionals, and civil servants who received education higher than the high school level. They were exempted from performing military service according to the 1933 compulsory conscription law issued by the GMD. The rule on their exemption was cancelled in the 1943 revised conscription law.

48. For the Western definition of civil society and public sphere and the debate of its existence in late Qing and Republican China, see *Modern China* 19, no. 2 (Symposium: "Public Sphere"/"Civil Society" in China? Paradigmatic Issues in Chinese Studies) (April 1993): 107–240; Prasenjit Duara, *Rescuing History from the Nation: Questioning Narratives of Modern China* (Chicago: Univ. of Chicago Press, 1995), 147–175; Heath Chamberlain, "Civil Society with Chinese Characteristics," *China Journal* 39 (January 1998): 69–81.

49. Bryna Goodman, "Creating Civic Ground: Public Maneuverings and the State in the Nanjing Decade," in *Remapping China: Fissures in Historical Terrain,* ed. Gail Hershatter et al. (Stanford, Calif.: Stanford Univ. Press, 1996), 164.

50. Ibid., 165. Also see Bryna Goodman, *Native Place, City, and Nation: Regional Networks and Identities in Shanghai, 1853–1937* (Berkeley: Univ. of California Press, 1995).

51. Xu, *Chinese Professionals and the Republican State,* 15.

52. Prasenjit Duara, *Culture, Power and the State: Rural North China, 1900–1942* (Stanford, Calif.: Stanford Univ. Press, 1988), 2.

53. Eastman, "Nationalist China during the Sino-Japanese War, 1937–1945," 127.

54. I translate *jingshen jiaoyu* as "civic education" instead of "ideological education" and "ideological indoctrination." This translation is intended to highlight that the Whampoa Military Academy aimed to train the cadets to be model citizen-soldiers who followed the code of ethics forged in Chiang Kai-shek's speeches and lectures at Whampoa.

55. The Three Principles of the People is a political philosophy developed by

Sun Yat-sen (1866–1925), who is referred to as the "Father of the Nation" in Republican China. This philosophy is often translated into nationalism, democracy, and the livelihood of the people. It has been claimed as the cornerstone of the Republic of China's policy as carried by the Chinese Nationalist party (Guomindang).

56. Susan Glosser, *Chinese Visions of Family and State, 1915–1953* (Berkeley: Univ. of California Press, 2003), 19.

57. Eugenia Lean, *Public Passions: The Trial of Shi Jianqiao and the Rise of Popular Sympathy in Republican China* (Berkeley: Univ. of California Press, 2007).

58. The Youth Army is used in this book as an abbreviation of the Educated Youth Expeditionary Army (*Zhishi qingnian yuanzhengjun*).

59. Lean, *Public Passions*, 13.

60. Ibid.

61. David Graff and Robin Higham, "Introduction," in *A Military History of China*, ed. David Graff and Robin Higham (Lexington: Univ. Press of Kentucky, 2012), 1–2.

62. Several books relate to the topics of Chinese military armament and the role of foreign advisers: Guangqiu Xu, *War Wings: The United States and Chinese Military Aviation, 1929–1949* (Westport, Conn.: Greenwood Press, 2001); William Leary, *The Dragon's Wing: The China National Aviation Corporation and Development of Commercial Aviation in China* (Athens: Univ. of Georgia Press, 1976); Daniel Ford, *Flying Tigers: Claire Chennault and the American Volunteer Group* (Washington, D.C.: Smithsonian Institution Press, 1991); William Kirby, *Germany and Republican China* (Stanford, Calif.: Stanford Univ. Press, 1984).

63. The typical studies on warlord armies written in the 1960s, 1970s, and 1980s are James Sheridan, *Chinese Warlord: The Career of Feng Yu-hsiang* (Stanford, Calif.: Stanford Univ. Press, 1966); Donald Gillin, *Warlord: Yen Hsi-shan in Shansi Province, 1991–1949* (Princeton, N.J.: Princeton Univ. Press, 1970); Diana Lary, *Region and Nation: The Kwangsi Clique in Chinese Politics, 1925–1937* (Cambridge, UK: Cambridge Univ. Press, 1974); Chi Hsi-sheng, *Warlord Politics in China, 1916–1928* (Stanford, Calif.: Stanford Univ. Press, 1976); Gavan McCormack, *Chang Tso-lin in Northeast China, 1911–1928* (Folkestone, UK: Dawson and Sons, 1977); Odoric Y. K. Wou, *Militarism in Modern China: The Career of Wu Pei-fu* (Folkestone, UK: Dawson and Sons, 1978); and Donald Sutton, *Provincial Militarism and the Chinese Republic: The Yunnan Army* (Ann Arbor: Univ. of Michigan Press, 1980). For more recent biographies of warlords in Republican China, see David Bonavia, *China's Warlords* (Hong Kong: Oxford Univ. Press, 1995). For general surveys of warlord and GMD armies, including the uniforms, equipment, and weaponry, see Philip Jowett's three books: *Chinese Civil War Armies, 1911–1949* (Oxford: Osprey Publishing, 1997); *The Chinese Army, 1937–1949: World War II and Civil War* (Oxford: Osprey Publishing, 2005); and *Chinese Warlord Armies, 1911–1930* (Oxford: Osprey Publishing, 2010). These three books contain substantial color illustration and serve the needs of broader audiences interested in military history.

64. Hans van de Ven, "The Military in the Republic," *China Quarterly* 150 (June 1997): 358.

65. Lary, *Warlord Soldiers*, 51.

66. Several scholars argue that the Nationalist army fought bravely during the Second Sino-Japanese War. For example, James Hsiung argues that Nationalist forces fought in all the major positional battles in the Second Sino-Japanese War. See James Hsiung, "The War and After: World Politics in Historical Perspective," in *China's Bitter Victory: The War with Japan, 1937–1941*, ed. James Hsiung and Steven Levine (Armonk, N.Y.: M. E. Sharpe, 1992), 295–306. Chang Jui-te argues that the Nationalist officer corps, although originally poor, improved through the period of the Second Sino-Japanese War with better professional training. See Chang Jui-te, "Nationalist Army Officers during the Second Sino-Japanese War, 1937–1945," *Modern Asian Studies* 30, no. 4 (October 1996): 1033–1056.

67. Arthur Waldron, *From War to Nationalism: China's Turning Point* (Cambridge, UK: Cambridge Univ. Press, 1995).

68. Eugene Levich, *The Kwangsi Way in Kuomintang China, 1931–1939* (Armonk, N.Y.: M. E. Sharpe, 1993).

69. Timothy Brook, "Preface: Lisbon, Xuzhou, Auschwitz: Suffering as History," in *Beyond Suffering: Recounting War in Modern China*, ed. James Flath and Norman Smith (Vancouver, Canada: Univ. of British Columbia Press, 2011), xii.

70. Influential studies on the rise of the CCP are Chalmers Johnson, *Peasant Nationalism and Communist Power: The Emergence of Revolutionary China* (Stanford, Calif.: Stanford Univ. Press, 1962); Yung-fa Chen, *Making Revolution: The Communist Movement in Eastern and Central China, 1937–1945* (Berkeley: Univ. of California Press, 1986); and Mark Selden, *China in Revolution: The Yenan Way Revisited* (New York: M. E. Sharpe, 1995).

71. Joseph Schumpeter, *Capitalism, Socialism, and Democracy* (New York: Harper and Row, 1975), 5; Patricia Stranahan, *Underground: The Shanghai Communist Party and the Politics of Survival, 1927–1937* (Lanham, Md.: Rowman and Littlefield, 1998), 16; Joseph Esherick, "Ten Theses on the Chinese Revolution," *Modern China* 21, no. 1 (January 1995): 47; Benjamin Schwartz, *Chinese Communism and the Rise of Mao* (Cambridge, Mass.: Harvard Univ. Press, 1951), 5.

72. Hung-Yok Ip, *Intellectuals in Revolutionary China: Leaders, Heroes and Sophisticates* (London: Routledge, 2005), 7.

73. Ibid., 7.

74. Elizabeth Perry, "Moving the Masses: Emotion Work in the Chinese Revolution," *Mobilization: An International Quarterly* 7, no. 2 (summer 2002): 111–128.

75. Yu Liu, "Maoist Discourse and the Mobilization of Emotions in Revolutionary China," *Modern China* 36, no. 3 (May 2010): 329–362.

76. Lean, *Public Passions*.

77. The *yangge* has been closely examined by David Holm's study *Art and Ideology in Revolutionary China*. Holm shows how the *yangge* underwent dramatic political transformations in the hands of the Communist propagandists in their

attempt to realize the Maoist goal of facing the masses. See David Holm, *Art and Ideology in Revolutionary China* (Oxford: Clarendon Press, 1991). What is absent in Holm's work is a systematic analysis of the *yangge* themes. This present research will distill a crucial topic out of the wide array of plays discussed in Holm's study—the image of soldier heroes and the army-people relationship.

78. Wang, *When "I" Was Born,* 4.

79. Ibid., 8.

80. Another category of sources that is potentially useful for this book is the handwritten diaries of Chiang Kai-shek, presently housed at the Hoover Institution. The diaries of Chiang currently available cover the period from 1917, when Chiang was rising to power, to 1945, when China won the Second Sino-Japanese War. This book does not rely on these copies of Chiang's diaries for several reasons. Substantial portions of the documents are water damaged, stuck together, or missing entirely, reflecting the fragility and poor condition of the originals. In addition, Chiang's family members chose to keep some passages with Chiang's personal commentary private. For these reasons as well as due to limited time and space, Chiang's diaries were only consulted in researching military education among students.

1. Training Model Soldiers at the Whampoa Military Academy

1. Whampoa recruited female cadets in 1926 for the first time in Chinese history. But the number of woman cadets was very small; for this reason, this chapter does not discuss the discourse covering female cadets. Chapter 4 will examine Whampoa's expectations for female cadets and how the female cadet Xie Bingying represented her image in her autobiographies.

2. Chang Rui-te, "The National Army from Whampoa to 1949," in *A Military History of China,* ed. David Graff and Robin Higham (Lexington: Univ. Press of Kentucky, 2012), 196.

3. The academy settled in Kaohsiung after it moved to Taiwan.

4. Wen Wen, preface to *Guomindang zhongyang lujun xuexiao yu junshi zhuanke xuexiao* (The Nationalist Party's Central Army Officer Academy and colleges for military specialties), ed. Wen Wen (Beijing: Zhongguo wenshi chubanshe, 2010), 1.

5. Hans van de Ven, "The Military in the Republic," *China Quarterly* 150 (June 1997): 353.

6. Colin Green, "Turning Bad Iron into Polished Steel: Whampoa and the Rehabilitation of the Chinese Soldier," in *Beyond Suffering: Recounting War in Modern China,* ed. James Flath and Norman Smith (Vancouver, Canada: Univ. of British Columbia Press, 2011), 154.

7. Wang Ke-wen, *Modern China: An Encyclopedia of History, Culture, and Nationalism* (New York: Garland, 1998), 385.

8. Zhongguo Guomindang zhongyang weiyuanhui dangshi weiyuanhui, ed., *Guofu quanji* (Complete compilation of the works by the Father of Republic of

China Sun Yat-sen), vol. 2 (Taibei: Zhongguo Guomindang zhongyang weiyuan-hui dangshi weiyuanhui, 1981), 692.

9. Chang, "The National Army from Whampoa to 1949," 195.

10. Zhongguo shehui kexueyuan jindaishisuo, ed., *Sun Yat-sen quanji* (Complete Compilation of Sun Yat-sen), vol. 10 (Beijing: Zhonghua shuju, 1986), 300.

11. "Lujun junguan xuexiao kaoxuan xuesheng jianzhang" (Memorandum on testing and selecting cadets for the army officer school), in *Zhongguo Guomindang zhoukan* (Weekly newspaper of Chinese Nationalist Party) 10 (March 2, 1924), reprinted from Chen Ningsheng, *Jiang Jieshi he Huangpu xi* (Chiang Kai-shek and his Whampoa clique) (Zhengzhou: Henan renmin chubanshe, 1994), 27–28.

12. Green, "Turning Bad Iron into Polished Steel," 155.

13. Ibid. As John K. Fairbank notes, "politically indoctrinated armies were a new thing in modern China." See John K. Fairbank, *The United States and China* (Cambridge, Mass.: Harvard Univ. Press, 1983), 292.

14. Nearly every province in China established military academies from the 1880s onward. For example, the Baoding Army Officer Academy (*Baoding lujun junguan xuexiao*) was a military academy based in Baoding, China. It was initiated by Yuan Shikai (1859–1916) in 1902 to train officers for his New Army (*Xin-jun*)—the modernized Qing army corps founded in 1895. It closed in 1923. The Yunnan Martial School (*Yunnan jiangwutang*) was founded in 1909 by the late Qing (1644–1911) government to train the New Army. After the 1911 revolution it became a military academy controlled by the Yunnan warlord Long Yun (1884–1962). After the outbreak of the Second Sino-Japanese War, it became the Fifth Branch Campus of the Central Army Officer Academy in 1938. It closed in 1945.

15. Green, "Turning Bad Iron into Polished Steel," 155.

16. Luo Derong, preface to *Xinbian junren jingshen jiaoyu* (New edition of civic education for soldiers) (N.p.: Junweihui tewutuan zhengxunchu xianbing jiaodao zongdui zhengxunchu, 1932), 2.

17. Sun Tzu, *Sunzi bingfa* (The art of war), trans. Lionel Giles (El Paso, Tex.: El Paso Norte Press, 2005), 33.

18. Chiang Kai-shek, "Sunzi bingfa yu gudai zuozhan yuanze yiji jinri zhan-zheng yishuhuade yiyi zhi chanming" (An illustration of *Sunzi bingfa,* traditional warfare strategies and the artistic principles of modern warfare) [1925], in *Xian zongtong Jiang gong sixiang yanlun zongji* (Comprehensive collections of the thoughts and speeches of deceased President Chiang), vol. 25, ed. Qin Xiaoyi (Taibei: Zhongyang dangshi weiyuanhui, 1984), 273.

19. Shen Zhenchuan's birth and death years are unclear.

20. Shen Zhenchuan, "Wuhan fenxiao dibaqi xuesheng kangri shiwei shijian qianhou" (A demonstration against the Japanese by the cadets in the eighth class of the Wuhan Branch Campus of the Central Army Officer Academy) [1964], in *Guomindang zhongyang lujun xuexiao yu junshi zhuanke xuexiao* (The Nationalist Party's Central Army Officer Academy and colleges of military specialties), ed. Wen Wen (Beijing: Zhongguo wenshi chubanshe, 2010), 60.

21. Personal behavior of Whampoa cadets mainly refers to the daily routine at Whampoa, such as eating and hygiene.

22. Chiang Kai-shek, "Keku nailao yu kangkai xisheng zhi biyao" (Necessity of diligence, endurance, and heroic sacrifice) [1924], in *Huangpu xunlianji xuanji* (Selections from the collection of the training materials at the Whampoa Military Academy), ed. Guomin zhengfu junshi weiyuanhui zhengzhibu (N.p.: Guomin zhengfu junshi weiyuanhui zhengzhibu, 1938), 35.

23. Chen Yuhuan, *Huangpu junxiao diyiqisheng yanjiu* (Studies on the cadets in the first class of the Whampoa Military Academy) (Guangzhou: Zhongshan daxue chubanshe, 2007), 5–71.

24. Ibid., 141.

25. Cheng Ling and Liu Mengxin, "1924–1927 nian Huangpu junxiao jiaoyude xiandaixing tezheng" (The characteristics of modernity in the education at the Whampoa Military Academy between 1924 and 1927), in *Huangpu junxiao yanjiu* (Studies on the Whampoa Military Academy), vol. 3, ed. Shu Yang (Guangzhou: Zhongshan daxue chubanshe, 2008), 72.

26. Chiang Kai-shek, "Zhuzhong weisheng yu jingshen dikangde daoli" (The significance of sanitation and spiritual resistance) [1924], in *Huangpu xunlianji xuanji* (Selections from the collection of the training materials at the Whampoa Military Academy), ed. Guomin zhengfu junshi weiyuanhui zhengzhibu (N.p.: Guomin zhengfu junshi weiyuanhui zhengzhibu, 1938), 42.

27. Chiang Kai-shek, "Qingjie jiancha jiangping" (Speeches and comments on the examination of sanitation) [1924], in *Huangpu xunlianji diyiji jingshen xunlian* (Volume one on civic education in the training materials at the Whampoa Military Academy) (N.p.: N.p., [1925]), 266.

28. Chiang, "Zhuzhong weisheng yu jingshen dikangde daoli" [1924], 43.

29. Chiang Kai-shek, "Junshi jiaoyu zhi yaozhi yu junji zhi genyuan" (The principles of military education and the sources of military discipline) [1924], in *Huangpu xunlianji diyiji jingshen xunlian* (Volume one on civic education in the training materials at the Whampoa Military Academy) (N.p.: N.p., [1925]), 407.

30. Wang Zhuochao, "Yi Nanjing zhongyang junxiao" (Remembering the Central Army Officer Academy at Nanjing) [1982], in *Guomindang zhongyang lujun xuexiao yu junshi zhuanke xuexiao* (The Nationalist Party's Central Army Officer Academy and colleges of military specialties), ed. Wen Wen (Beijing: Zhongguo wenshi chubanshe, 2010), 4.

31. The historian Seungsook Moon Sebesta, although she studies the militarized modernity in South Korea, demonstrates the significance of daily routine in strengthening military discipline. Sebesta shows that "the routine aspects of the military subculture indicate that popular acceptance of military service as men's national duty was not grounded in any genetic inclination of males to violence but stemmed rather from a cultural inclination to obedience that would permit a man's integration into the highly hierarchical military system." See Seungsook

Moon Sebesta, *Militarized Modernity and Gendered Citizenship in South Korea* (Durham, N.C.: Duke Univ. Press, 2005), 39.

32. Chiang, "Qingjie jiancha jiangping" [1924], 266.

33. Chiang Kai-shek, "Chongzheng dongjiang xunjie" (Lectures on the second military campaign at East River) [1925], in *Huangpu xunlianji diyiji jingshen xunlian* (Volume one on civic education in the training materials at the Whampoa Military Academy) (N.p.: N.p., [1925]), 450.

34. Chiang Kai-shek, "Geming junrende renge" (Moral qualities of a revolutionary army soldier) [1924], in *Huangpu xunlianji diyiji jingshen xunlian* (Volume one on civic education in the training materials at the Whampoa Military Academy) (N.p.: N.p., [1925]), 287.

35. Chiang Kai-shek, "Xiaozhang diwuci xunci" (The fifth admonition by the commandant) [1924], in *Huangpu xunlianji diyiji jingshen xunlian* (Volume one on civic education in the training materials at the Whampoa Military Academy) (N.p.: N.p., [1925]), 32.

36. Leng Xin, "Huangpu junxiao wushi zhounian jinian ganxiang" (Thoughts on the fiftieth anniversary of the foundation of the Whampoa Military Academy [1974]), in *Zhanshi lunji* (Collection of the studies on the history of wars), ed. Wei Rulin (Taibei: Huagang chuban youxian gongsi, 1976), 58.

37. Green, "Turning Bad Iron into Polished Steel," 172.

38. Ibid.

39. Chiang Kai-shek, "Qiangde xingzhi yu zuoyong he junren naqiang zhi mudi" (The nature and functions of the rifle and the purpose of holding the rifle for a soldier) [1924], in *Huangpu xunlianji xuanji* (Selections from the collection of the training materials at the Whampoa Military Academy), ed. Guomin zhengfu junshi weiyuanhui zhengzhibu (N.p.: Guomin zhengfu junshi weiyuanhui zhengzhibu, 1938), 49–50.

40. Liang Tianli was a cadet in the third class of Whampoa Military Academy; his birth and death years are unclear.

41. Chiang Kai-shek, "Zisha shi beiqiede fanzui xingwei" (Suicide is a cowardly crime) [1925], in *Huangpu xunlianji diyiji jingshen xunlian* (Volume one on civic education in the training materials at the Whampoa Military Academy) (N.p.: N.p., [1925]), 469.

42. Ibid.

43. Chiang Kai-shek, "Benxiao jiaoyude fangzhen zhi yan zhi biyao" (Necessity of being strict as an educational principle at this academy) [1924], in *Huangpu xunlianji xuanji* (Selections from the collection of the training materials at the Whampoa Military Academy), ed. Guomin zhengfu junshi weiyuanhui zhengzhibu (N.p.: Guomin zhengfu junshi weiyuanhui zhengzhibu, 1938), 76. The term *wenzhi binbin* first appears in the *Yongye* passage from *The Analects,* a collection of the Chinese philosopher Confucius's sayings brought together by his pupils shortly after his death in 497 B.C. The term originally meant "elegant and refined in manner," but Chiang Kai-shek uses it in his speech to depict the image of a gentle

and fragile scholar. The term *jiujiu wufu* first appears in the "Tuzhi" poem praising a prince's warriors from *Shijing* (Book of odes), the oldest existing collection of Chinese poetry, comprising 305 works dating from the eleventh to the seventh century B.C.

44. Ibid.

45. Joshua Goldstein, *War and Gender: How Gender Shapes the War System and Vice Versa* (New York: Cambridge Univ. Press, 2001), 266.

46. Chiang, "Keku nailao yu kangkai xisheng zhi biyao" [1924], 39.

47. The years of Cheng Ruji's birth and death are unclear.

48. Chiang, "Keku nailao yu kangkai xisheng zhi biyao" [1924], 39.

49. The number of cadets belonging to the first class at Whampoa is controversial. For this controversy, see Chen, *Huangpu junxiao diyiqisheng yanjiu*, 2–3. This book adopts Chen's conclusion of 706.

50. Chen, *Huangpu junxiao diyiqisheng yanjiu*, 224.

51. William Kirby, *Germany and Republican China* (Stanford, Calif.: Stanford Univ. Press, 1984), 158.

52. Frederic Wakeman, *Spymaster: Dai Li and the Chinese Secret Service* (Berkeley: Univ. of California Press, 2003), 27.

53. Zhao Zhen's birth and death years are unclear.

54. Zhao Zhen, "Huiyi Jiang Jieshi dui zhongyang junxiao xueshengde longluo shouduan" (Memories on the techniques employed by Chiang Kai-shek to win over the cadets at the Central Army Officer Academy) [1964], in *Guomindang zhongyang lujun xuexiao yu junshi zhuanke xuexiao* (The Nationalist Party's Central Army Officer Academy and colleges of military specialties), ed. Wen Wen (Beijing: Zhongguo wenshi chubanshe, 2010), 14.

55. Xie Yingbai's birth and death years are unclear.

56. Mao Yuli's birth and death years are unclear.

57. Xie Yingbai, "1929 zhi 1933 nian de Nanjing zhongyang junxiao" (The Central Army Officer Academy at Nanjing between 1929 and 1933) [1963], in *Guomindang zhongyang lujun xuexiao yu junshi zhuanke xuexiao* (The Nationalist Party's Central Army Officer Academy and colleges of military specialties), ed. Wen Wen (Beijing: Zhongguo wenshi chubanshe, 2010), 12.

58. Zhao, "Huiyi Jiang Jieshi dui zhongyang junxiao xueshengde longluo shouduan" [1964], 14.

59. Ibid.

60. Ibid.

61. Qi Xiangming's birth and death years are unclear. Qi studied at Whampoa in 1930.

62. Qi Xiangming, "Jiuyiba qianhoude zhongyang junxiao" (The Central Army Officer Academy around the period of the Mukden Incident on September 18, 1931) [1963], in *Guomindang zhongyang lujun xuexiao yu junshi zhuanke xuexiao* (The Nationalist Party's Central Army Officer Academy and colleges of military specialties), ed. Wen Wen (Beijing: Zhongguo wenshi chubanshe, 2010), 18.

63. Andrew Scobell, *China's Use of Military Force: Beyond the Great Wall and the Long March* (Cambridge, UK: Cambridge Univ. Press, 2003), 62.

64. Chiang Kai-shek, "Gemingjun Lianzuofa" (Law of joint responsibility in the revolutionary army) [1925], in *Huangpu xunlianji* (Selections of the training materials at the Whampoa Military Academy), ed. Deng Wenyi (Wuhan: Guofangbu xinwenju, 1938), 345–346.

65. Du Congrong, *Huangpu junxiao zhi chuangjian ji dongzheng beifa zhi huiyi* (Memories of the foundation of the Whampoa Military Academy and the Eastern and Northern Expeditions) (Taibei: Shunren caise yinzhi youxian gongsi, 1975), 43.

66. Chiang Kai-shek, "Benxiao zhi shiming yu gemingde rensheng" (The mission of this academy and the revolutionary life) [1924], in *Huangpu xunlianji xuanji* (Selections from the collection of the training materials at the Whampoa Military Academy), ed. Guomin zhengfu junshi weiyuanhui zhengzhibu (N.p.: Guomin zhengfu junshi weiyuanhui zhengzhibu, 1938), 7.

67. Rebecca Nedostup, "Civic Faith and Hybrid Ritual in Nationalist China," in *Converting Cultures: Religion, Ideology, and Transformations of Modernity,* ed. Dennis Washburn and A. Kevin Reinhart (Leiden, Netherlands: Koninklijke Brill NV, 2007), 45.

68. Ming K. Chan and Arif Dirlik, *Schools into Fields and Factories: Anarchists, the Guomindang, and the National Labor University in Shanghai, 1927–1932* (Durham, N.C.: Duke Univ. Press, 1991), 103.

69. "Zunxing zongli jinian zhou yishi," in *Zhongyang lujun junguan xuexiao shigao* (History of the Central Army Officer Academy), vol. 7, ed. Guomindang zhongyang lujun junguan xuexiao xiaowu (Nanjing: Guomindang zhongyang lujun junguan xuexiao xiaowu, 1936), 22.

70. Chan and Dirlik, *Schools into Fields and Factories,* 103.

71. The weekly memorial service did not go smoothly. The Nanjing government, for instance, was forced to amend the ceremony's rules to cover such matters as punishments for absenteeism. See Nedostup, "Civic Faith and Hybrid Ritual in Nationalist China," 45.

72. Li Ning, "Huangpu junxiao jinianri he liyi yu jidujiao guanxi chutan" (An initial study on the relations of the holidays and rituals observed at the Whampoa Military Academy with Christianity), in *Huangpu junxiao yanjiu* (Studies on the Whampoa Military Academy), vol. 3, ed. Shu Yang (Guangzhou: Zhongshan daxue chubanshe, 2008), 92.

73. Patricia Ebrey, *Confucianism and Family Rituals in Imperial China: A Social History of Writing about Rites* (Princeton, N.J.: Princeton Univ. Press, 1991), 14.

74. Chen, *Huangpu junxiao diyiqisheng yanjiu,* 192.

75. Ibid.

76. Ibid., 193.

77. Ibid.

78. Qi, "Jiuyiba qianhoude zhongyang junxiao" [1963], 16.

79. Scobell, *China's Use of Military Force,* 61.

80. Hans van de Ven, *War and Nationalism in China, 1925–1945* (New York: Routledge Curzon, 2003), 151.

81. Xie, "1929 zhi 1933 nian de Nanjing zhongyang junxiao" [1963], 8.

82. Ibid.

83. Li Yihu's birth and death years are unclear.

84. Zhang Zhizhong commanded the Fifth Army of the Nationalist military forces in the 1932 Battle of Shanghai against Japan. Later, as the head of the Ninth Army Group of the Nationalist forces, Zhang supervised the defense of Shanghai against Japan in 1937.

85. Xie, "1929 zhi 1933 nian de Nanjing zhongyang junxiao" [1963], 8.

86. Lincoln Li, *Student Nationalism in China, 1924–1949* (Albany: State Univ. of New York Press, 1994), 48.

87. Chen Tingrong's birth and death years are unclear.

88. Cheng Tingrong, "Chengdu zhongyang junxiao xuechao jishi" (The cadet uprising in the Chengdu Branch Campus of the Central Army Officer Academy) [1963], in *Guomindang zhongyang lujun xuexiao yu junshi zhuanke xuexiao* (The Nationalist Party's Central Army Officer Academy and colleges of military specialties), ed. Wen Wen (Beijing: Zhongguo wenshi chubanshe, 2010), 57.

89. Tan Dingyuan, "Zhongyang junxiao shi'erqi jianwen" (Experiences in the twelfth class of the Central Army Officer Academy) [1961], in *Guomindang zhongyang lujun xuexiao yu junshi zhuanke xuexiao* (The Nationalist Party's Central Army Officer Academy and colleges of military specialties), ed. Wen Wen (Beijing: Zhongguo wenshi chubanshe, 2010), 22.

90. Ibid., 25. The first character of the Chinese word *huangchong* is pronounced the same as the first character of the Chinese word *Huangpu* (formerly romanized as Whampoa).

91. Ibid.

92. Frederick Fu Liu, *A Military History of Modern China* (Princeton, N.J.: Princeton Univ. Press, 1956).

93. Eugene Levich, *The Kwangsi Way in Kuomintang China, 1931–1939* (Armonk, N.Y.: M. E. Sharpe, 1993), 12.

94. Ibid.

95. Tan, "Zhongyang junxiao shi'erqi jianwen" [1961], 25.

96. Edward McCord, "Warlordism in Early Republican China," in *A Military History of China,* ed. David Graff and Robin Higham (Lexington: Univ. Press of Kentucky, 2012), 190.

97. Qian Daquan was a cadet from the eleventh class at Whampoa. He graduated in 1937.

98. Qian Daquan, "Zhongyang junxiao xiqian jishi" (Narratives on the relocation of the Central Army Officer Academy to the west) [1963], in *Guomindang zhongyang lujun xuexiao yu junshi zhuanke xuexiao* (The Nationalist Party's Central Army Officer Academy and colleges of military specialties), ed. Wen Wen (Beijing: Zhongguo wenshi chubanshe, 2010), 38–39.

99. Zheng Dianqi, "Luoyang fenxiao yu junguan xunlianban" (The Luoyang branch campus and the military officer training class) [1963], in *Guomindang zhongyang lujun xuexiao yu junshi zhuanke xuexiao* (The Nationalist Party's Central Army Officer Academy and colleges of military specialties), ed. Wen Wen (Beijing: Zhongguo wenshi chubanshe, 2010), 74.

100. Chang, "The National Army from Whampoa to 1949," 197.

101. Ibid.

102. Ibid.

2. Enlisting Citizens in the Military Mobilization of the Nationalist State

1. David Graff and Robin Higham, "Introduction," in *A Military History of China*, ed. David Graff and Robin Higham (Lexington: Univ. Press of Kentucky, 2012), 10.

2. Diana Lary, *Warlord Soldiers: Chinese Common Soldiers, 1911–1937* (Cambridge, UK: Cambridge Univ. Press, 1985), 14.

3. Wang Xiaogui and Gong Zeqi, *Zhongguo jindai junren daiyu shi* (History of the treatment of soldiers in modern China) (Beijing: Haichao chubanshe, 2006), 316.

4. Hans van de Ven, "New States of War: Communist and Nationalist Warfare and State Building (1928–1934)," in *Warfare in Chinese History*, ed. Hans van de Ven (Leiden, Netherlands: Koninklijke Brill NV, 2000), 323.

5. Colin Green, "Turning Bad Iron into Polished Steel: Whampoa and the Rehabilitation of the Chinese Soldier," in *Beyond Suffering: Recounting War in Modern China*, ed. James Flath and Norman Smith (Vancouver, Canada: Univ. of British Columbia Press, 2011), 155.

6. Lary, *Warlord Soldiers*, 16.

7. Xu Chonghao, *Zhengbing zhi yange ji shixingfa* (The transformation of the military recruiting system and its implementation) (Nanjing: Minzhi shuju, 1929), 2–6.

8. Hans van de Ven, *War and Nationalism in China, 1925–1945* (New York: Routledge Curzon, 2003), 143.

9. Ibid., 144.

10. Jae Ho Chung, "The Evolving Hierarchy of China's Local Administration: Traditions and Changes," in *China's Local Administration: Traditions and Changes in the Sub-National*, ed. Jae Ho Chung and Tao-chiu Lam (New York: Routledge, 2010), 7.

11. Ibid.

12. Chung, "The Evolving Hierarchy of China's Local Administration," 7.

13. James Zheng Gao, *Historical Dictionary of Modern China (1800–1949)* (Lanham, Md.: Scarecrow Press, 2009), 20.

14. Van de Ven, "New States of War," 357.

15. Graff and Higham, "Introduction," 10.

16. Zhongyang xunliantuan bingyi ganbu xunlianban, ed., *Bingyi fagui huibian, yiwu* (Compilation of laws and regulations on conscription, drafting affairs) (N.p.: N.p., 1942), 55.

17. Van de Ven, *War and Nationalism in China*, 145.

18. Ibid.

19. Ibid., 146.

20. According to Sun Yat-sen's plans, the GMD was to rebuild China in three steps: military rule (*junzheng*), political tutelage (*xunzheng*), and constitutional rule (*xianzheng*). After the National Revolutionary Army ended its northern campaigns in 1928, China came under the political tutelage of the GMD. The principal aim of this period was to enable the party, which considered itself to be the elite of the Chinese nation, "to instruct the people in the pursuit of constructive work of a revolutionary nature." The organization of a new government in 1947, according to the Constitution adopted on December 25, 1946, formally terminated the period of political tutelage. However, in reality, there was little difference between the two periods of political tutelage and constitutional rule. See Tuan-sheng Chien, *The Government and Politics of China, 1912–1949* (Stanford, Calif.: Stanford Univ. Press, 1950), 134.

21. Jin Mingsheng, *Zhonghua minguo xianfa cao'an shiyi* (Illustration on the Republic of China Constitution draft) (Shanghai: Shijie shuju, 1936), 42.

22. Junzhengbu bingyishu yizhengsi xuanchuanbu, ed., *Bingyi xuanchuan ji youdai zhengshu faling huibian* (Compilation of conscription propaganda as well as laws and regulations on favorably treating soldiers' dependents) (Chongqing: Junzhengbu bingyishu yizhengsi xuanchuanbu, 1943), 39.

23. Zhongyang xunliantuan bingyi ganbu xunlianban, ed., *Bingyi fagui huibian, yiwu*, 57.

24. Ibid., 68.

25. Jay Taylor, *The Generalissimo's Son: Chiang Ching-kuo and the Revolutions in China and Taiwan* (Cambridge, Mass.: Harvard Univ. Press, 2000), 119.

26. Wang and Gong, *Zhongguo jindai junren daiyu shi*, 317.

27. Ibid.

28. Another big change that the 1943 revised conscription law made to the 1933 law was the deletion of the rule that students in senior and higher-level schools were exempted from performing military service. The impact of this change on the discourse of the soldier figure is discussed in chapter 5.

29. Junzhengbu bingyishu yizhengsi xuanchuanbu, ed., *Bingyi xuanchuan ji youdai zhengshu faling huibian*, 8.

30. Ibid., 45.

31. Shi Mei et al., eds., *Kangzhan jianguo gangling wenda* (Questions and answers on the guidelines for resisting the Japanese and building the state) (Shanghai: Shenghuo shudian, 1938), 3.

32. Zhongyang xunliantuan bingyi ganbu xunlianban, ed., *Bingyi fagui huibian, yiwu*, 285.

33. Junzhengbu bingyishu yizhengsi xuanchuanbu, ed., *Bingyi xuanchuan ji youdai zhengshu faling huibian,* 8.

34. Ibid.

35. Zhongyang xunliantuan bingyi ganbu xunlianban, ed., *Bingyi fagui huibian, yiwu,* 289.

36. Junzhengbu bingyishu yizhengsi xuanchuanbu, ed., *Bingyi xuanchuan ji youdai zhengshu faling huibian,* 8.

37. Arthur Waldron, "The Warlord: Twentieth-Century Chinese Understanding of Violence, Militarism, and Imperialism," *American Historical Review* 96, no. 4 (1991): 1073–1100.

38. The military theorist General Sir John Hackett analyzes the concept that the military is treated in many states as a "repository of moral resource." See General Sir John Hackett, "The Military in the Service of the State," in *War, Morality, and the Military Professor,* ed. Malham Wakin (Boulder, Colo.: Westview Press, 1986), 119.

39. Zhu Peide served as the general inspector (*zongjian*) at the General Inspector Branch (zongjian bu) of the Nationalist army during the war.

40. Zhu Peide, *Junguan de xinshenghuo* (New life of military officers) (Nanjing: Zhengzhong shuju, 1934), 11.

41. The National Revolutionary Army Reorganization Meeting (*Guomin gemingjun bianqian huiyi*) was held on January 1, 1929. It set up eight reorganized military zones (*bianqianqu*) within the nation: Central Reorganized Military Zone (led by Chiang Kai-shek), Navy Zone (led by Chiang), No. 1 Zone (led by Chiang), No. 2 Zone (led by Feng Yuxiang), No. 3 Zone (led by Yan Xishan), No. 4 Zone (led by Li Zongren), No. 5 Zone (led by Northeastern warlords), and No. 6 Zone (led by warlords in Sichuan, Yunnan, Guizhou, and Xikang).

42. Wang and Gong, *Zhongguo jindai junren daiyu shi,* 399.

43. Ibid. The Nineteenth Route Army was a Nationalist army led by General Cai Tingkai (1892–1968). It gained a good reputation among Chinese for fighting the Japanese in Shanghai in the 1932 Battle of Shanghai.

44. According to the "Haijun lianbing zhaomu jianzhang" (General regulations on recruiting navy forces) issued in October 1929, the new recruits in the navy force had to have a basic literacy education (*chushi wenzi*). See Zhongguo di'er lishi dang'an guan, ed., *Zhonghua minguoshi dang'an ziliao huibian* (Compilations of archives and materials during the Republic of China period), vol. 5, no. 1 (Nanjing: Jiangsu guji chubanshe, 1994), 266.

45. Wang and Gong, *Zhongguo jindai junren daiyu shi,* 400.

46. This literacy textbook for soldiers is the only one that is currently kept in the Shanghai Municipal Library.

47. Junshi weiyuanhui weiyuanzhang Nanchang xingying, ed., *Shibing shizi keben, disance* (Literacy textbooks for soldiers, level 3) (Shanghai: Zhonghua shuju, 1935), 1–4.

48. Ibid., 6–8, 18.

49. Ibid., 21–25.

50. Lloyd Eastman, "Nationalist China during the Sino-Japanese War, 1937–1945," in *The Nationalist Era in China, 1927–1949,* ed. Lloyd Eastman et al. (New York: Cambridge Univ. Press, 1991), 126.

51. Junshi weiyuanhui junxunbu, *Zhanshi lujun jiaoyuling cao'an* (The draft of education decree in wartime period) (Chongqing: Junshi weiyuanhui junxunbu, 1944), 1.

52. Van de Ven, *War and Nationalism,* 164.

53. Zhu, *Junguan de xinshenghuo,* 1.

54. Ibid.

55. Ibid., 2.

56. Ibid.

57. Ye Shoukang et al., *Junren shouce* (Handbooks for soldiers) (Jinhua: Zhejiang junxun tushu chubanshe, 1939), 33.

58. Ibid.

59. Zhongyang xunliantuan bingyi ganbu xunlianban, ed., *Bingyi fagui huibian, yiwu,* 124.

60. Junzhengbu bingyishu yizhengsi xuanchuanbu, ed., *Bingyi xuanchuan ji youdai zhengshu faling huibian,* 57.

61. Wang and Gong, *Zhongguo jindai junren daiyu shi,* 335. Corporal, sergeant, and staff sergeant were military titles of noncommissioned officers. Private E-1, private E-2, and private first class E-3 constituted the three rules of the enlisted personnel.

62. Tang Degang, *Li Zongren huiyi lu* (Memoirs of Li Zongren) [1958] (Hongkong: Nanyue chubanshe, 1986), 330.

63. Zhongguo di'er lishi dang'an guan, ed., *Zhonghua minguoshi dang'an ziliao huibian,* vol. 5, no. 1, 216.

64. Wang and Gong, *Zhongguo jindai junren daiyu shi,* 320.

65. Ibid., 336.

66. Van de Ven, *War and Nationalism in China,* 257.

67. The National Political Council Conference (*Guomin canzhenghui*) was a political agent under the Nationalist rule, which existed between July 1938 and March 1948.

68. Wang and Gong, *Zhongguo jindai junren daiyu shi,* 320.

69. R. J. Rummel, *Death by Government* (New Brunswick, N.J.: Transaction Publishers, 2009), 130. During the periods of the Nanjing decade and the Second Sino-Japanese War, the soldiers who received relatively good treatment were those in the Youth Army, which consisted of students and civil servants. The construction of the soldier figure by the students is discussed in chapter 5.

70. Lloyd Eastman, *Seeds of Destruction: Nationalist China in War and Revolution, 1937–1949* (Stanford, Calif: Stanford Univ. Press, 1984), 151.

71. Theodore H. White, *Thunder out of China* (New York: Da Capo Press, 1980).

72. Jack Belden, *China Shakes the World* (New York: Monthly Review Press, 1970).

73. Ibid., 147–148.

74. Ibid.

75. Van de Ven, *War and Nationalism in China,* 254.

3. Wartime Soldier Support by Urban Intellectuals and Professionals

1. Leo Ou-fan Lee, "Literary Trends: The Road to Revolution," in *An Intellectual History of Modern China,* ed. Merle Goldman and Leo Ou-fan Lee (New York: Cambridge Univ. Press, 2002), 242.

2. Ibid.

3. Hu Lanqi, *Zhandi ernian* (Two years at the front line) (Ji'an: Laodong funü zhandi fuwutuan, 1939), 1.

4. Liu Naifu, *Zhandi fuwu gongzuo yu jingyan* (Experiences in performing war area service work) (Hankou: Shenghuo shudian, 1938), 2.

5. The United Association of Various Circles for National Salvation was a social organization founded by patriotic intellectuals in Shanghai on May 31, 1936. Its members also included pro-democracy activists with no explicit party affiliation. Examples of its members are Tao Xingzhi (1891–1946), Zou Taofen (1895–1944), Shi Liangcai (1880–1934), and Song Qingling (1890–1981). For a detailed study on this association, see Parks Coble, "Chiang Kai-shek and the Anti-Japanese Movement in China: Zou Taofen and the National Salvation Association, 1931–1937," *Journal of Asian Studies* 44, no. 2 (February 1985): 293–310.

6. Tianjin's *Dagongbao* was the oldest active Chinese language newspaper in China. It was founded on June 17, 1902, by the Catholic Manchu publisher Ying Lianzhi (1867–1926) in Tianjin in order to realize a modern and democratic nation. It had a reformist agenda, namely "to increase the level of education of the people in order to change the evil customs of our country." It regularly published a *baihua* column and saw itself as "a forum to discuss methods of popular enlightenment." See Elisabeth Kaske, *The Politics of Language in Chinese Education, 1895–1919* (Leiden, Netherlands: Koninklijke Brill NV, 2008), 142. With "no party affiliation, no political endorsement, no self-promotion, no ignorance" as its motto, the newspaper's popularity quickly rose because of its sharp political commentary, especially of the Japanese as the Second Sino-Japanese War began. As the war continued, the journalists fled to other cities, such as Shanghai, Hankou, Chongqing, Guilin, and Hong Kong, to continue publishing.

7. *Jiuwang ribao* was the newspaper of the Literature and Art Circles National Salvation Association and was published by the CCP. See Patricia Stranahan, *Underground: The Shanghai Communist Party and the Politics of Survival, 1927–1937* (Lanham, Md.: Rowman and Littlefield, 1998), 220.

8. Lan Hai, *Zhongguo kangzhan wenyi shi* (A history of the literatures in China's Anti-Japanese War) (Jinan: Shandong wenyi chubanshe, 1984), 84.

9. Ibid.

10. Fan Changjiang, "Yi ye zhanchang" (Memories of the wild battlefield), in

Zhandi guilai (Return from the battlefield), ed. Tian Han (N.p.: Zhanshi chuban-she, [1938]), 40.

11. Ibid., 42.

12. Xie Bingying, *Xie Bingying daibiaozuo* (Representative works by Xie Bingy-ing) (Beijing: Huaxia chubanshe, 2009), 207.

13. Ibid.

14. Fan, "Yi ye zhanchang," 42.

15. Fan Changjiang, "Lugouqiao pan" (Along the Lugou Bridge), in *Fan Chang-jiang xinwen wenji* (Collections of Fan Changjiang's news reports), ed. Shen Pu (Beijing: Xinhua chubanshe, 2001), 618.

16. Li Mingjian, "Zhandi jianying" (The sketch of the battlefield), in *Zhandi guilai* (Return from the battlefield), ed. Tian Han (N.p.: Zhanshi chubanshe, [1938]), 63.

17. Wu Dakun, "Qianxian liang zhouye" (Two days and nights at the front), in *Zhandi guilai* (Return from the battlefield), ed. Tian Han (N.p.: Zhanshi chuban-she, [1938]), 69.

18. Shen Qiyu, "Qianxian guilai ji" (Return from the front line), in *Zhandi guilai* (Return from the battlefield), ed. Tian Han (N.p.: Zhanshi chubanshe, [1938]), 24.

19. Yang Fenjun's birth and death years are unknown.

20. Yang Fenjun, "Zai yedi yiyuan" (At the hospital in the front line), in *Zhandi guilai* (Return from the battlefield), ed. Tian Han (N.p.: Zhanshi chubanshe, [1938]), 88.

21. Fan, "Yi ye zhanchang," 40.

22. Ibid.

23. Ibid., 42.

24. Li, "Zhandi jianying," 62.

25. Wu, "Qianxian liang zhouye," 66, 77.

26. Liu Liangmo, "Zai zhandi yiyuan li" (At the hospital in the front line), in *Zhandi guilai* (Return from the battlefield), ed. Tian Han (N.p.: Zhanshi chuban-she, [1938]), 84.

27. Xie, *Xie Bingying daibiaozuo*, 181.

28. Wu, "Qianxian liang zhouye," 77.

29. Fan, "Yi ye zhanchang," 40, 42.

30. Wu, "Qianxian liang zhouye," 77.

31. Xie, *Xie Bingying daibiaozuo*, 182.

32. *Fenghuo* was a newspaper founded by Mao Dun (1896–1981) in 1937. It was originally named *Nahan* (Crying out). Its chief editor was Ba Jin (1904–2005).

33. Tian Zhongji used the pen name Lan Hai during the war.

34. Lan, *Zhongguo kangzhan wenyi shi*, 146.

35. Huizhu, "Zai shangbing yiyuan" (At the hospital for wounded soldiers), in *Zhandou de suhui* (Quick sketches of the combat), ed. Yi Qun (Chongqing: Zuojia shuwu, 1943), 31.

36. Ibid.

37. Ibid., 32.

38. Ibid., 33.

39. Ibid., 34.

40. Ibid., 35.

41. Ibid.

42. Ibid.

43. Xinshcnghuo yundong funü zhidaohui, *Rongyu junren fuwu gongzuo jishi* (Records of the work of serving honorable soldiers) (Chongqing: N.p., 1944), 39.

44. Ibid., 49.

45. Ibid.

46. Zhao Xiaoyang, "Kangri zhanzheng shiqi Zhongguo jidujiao qingnianhui junren fuwubu yanjiu" (The study on the Emergency Service to Soldiers Program of the Chinese YMCA during the Anti-Japanese War), *Kangri zhanzheng yanjiu* (Studies on the Anti-Japanese War) 2 (2011): 31. For a detailed study of the Emergency Service to Soldiers Program of the Chinese YMCA, please see Yan Xu, "Befriending Soldiers: The Emergency Service to Soldiers Program of the Chinese YMCA during the Second Sino-Japanese War, 1837–1945," in *The YMCA at War: Collaboration and Conflict during the World Wars,* ed. Jeffrey Copeland and Yan Xu (Lanham, Md.: Lexington Books, 2018): 161–176.

47. Liu Liangmo, "Zhanshide junren fuwu" (Soldier service in wartime) (Hankou: Xinzhi shudian, 1938), 31.

48. Ibid., 6.

49. Ibid., 32.

50. Ibid., 38.

51. Liang Xiaochu, *Zhonghua jidujiao qingnianhui zhanqu fuwu quanguo weiyuanhui baogaoshu* (Reports of the national committee for war area service led by the Chinese YMCA) (N.p.: Zhonghua jidujiao qingnianhui, 1933), 9–14.

52. Zhao Yamin, "Pianduan de huiyi: cong Shandong dao Yunnan" (Fragmented memories: From Shandong to Yunnan), in *Kangri jiuwang shiride lishi huigu* (The historical review of the years during the Anti-Japanese War), ed. Quanguo jidujiao qingnianhui junren fuwubu tonggong (Ha'erbin: Shuangchengshi yinchuachang, 1994), 56.

53. Ouyang Bo, "Chongchu Changsha dahuo, benxiang qiantangjiang bian" (Rushing out of the fire in Changsha, marching to the Qianjiang riverside," in *Kangri jiuwang shiride lishi huigu* (The historical review of the years during the Anti-Japanese War), ed. Quanguo jidujiao qingnianhui junren fuwubu tonggong (Ha'erbin: Shuangchengshi yinchuachang, 1994), 142.

54. Liang, *Zhonghua jidujiao qingnianhui zhanqu fuwu quanguo weiyuanhui baogaoshu,* 9.

55. Liu, "Zhanshide junren fuwu," 8.

56. Ibid.

57. Ibid.

58. Ibid., 42.

59. Ibid., 10.

60. Ibid., 12.

61. Ibid.

62. Ibid.

63. Ibid., 15.

64. Ibid., 19.

65. Ibid., 11.

66. Ibid., 34.

67. Ibid., 44.

68. Ibid.

69. Ibid., 36.

70. Ibid., 45.

71. Ibid., 46.

72. The semimonthly magazine *Shangbing zhi you* (Friends of the wounded soldiers) was edited and published by the Jiangxi Honorable Soldiers Administration Center (Jiangxi rongyu junren guanlichu).

73. Chiang Ching-kuo, "Gao fushang jiangshi shu—wei qingzhu sanshinian yuandan er zuo" (To the wounded soldiers—a speech written for the celebration of the new year of 1941), *Shangbing zhi you* (Friends of the wounded soldiers) 9 (January 1, 1941): 9.

74. Chen Junming's birth and death years are unknown.

75. Chen Junming, *Yige jinjide huyu: shangbing gongzuo zhi lilun, shiji yu jianyi* (An emergent call: The theory, practice, and suggestion for wounded soldier work) (Changsha: Zhongxin gongsi, 1939), 4.

76. Ibid., 31.

77. Peng Xiuliang, "Duan Shengwu" (Biography of Duan Shengwu), *Tuixiu shenghuo* (Retirement life) 12 (2010): 35.

78. Fu Sinian, "Duan Shengwu xiansheng zhuan" (Biography of Mr. Duan Shengwu), *Wenshi zazhi* (Journal of literature and history) 5, no. 7 (1945): 47.

79. Rongyu junren zhiye xiedaohui, ed., *Rongyu junren zhiye xiedaohui diyi niandu gongzuo baogao* (Reports on the first-year work of the Vocational Coordination Association for the Disabled Veterans) (Chongqing: Rongyu junren zhiye xiedaohui, 1941), 4.

80. Ibid., 7.

81. Ibid.

82. Gou Xingchao, "Kangzhan shiqide suican bufei yundong" (The campaign of disabled-but-not-useless during the Second Sino-Japanese War), *Wenshi zazhi* (Journal of literature and history) 5 (2007): 55.

83. Ibid.

84. Yu Zhaoming's birth and death years are unknown.

85. The Ministry of Society (*Shehuiju*) of the GMD government was estab-

lished in 1938 and subject to the leadership of the GMD Central Committee. It was mainly responsible for social welfare. Its first minister was Gu Zhenggang (1901–1993).

86. Yu Zhaoming, *Rongyu junren zhi zhiye zaizao* (Vocational rehabilitation for honorable soldiers) (N.p.: Junshi weiyuanhui houfang qinwu bu zhengzhi bu, 1942), 1.

87. Ibid., 17.

88. Ibid., 5.

89. Ibid.

90. Rongyu junren zhiye xiedaohui, ed., *Rongyu junren zhiye xiedaohui diyi niandu gongzuo baogao,* 15.

91. Yu Zhaoming, *Rongyu junren jiuye fudao* (Assisting honorable soldiers in getting employed) (Nanjing: Zhengzhong shuju, 1947), 49.

92. Chen, *Yige jinjide huyu,* 34.

93. Sichuansheng difangzhi bianzuan weiyuanhui, *Sichuansheng zhi, minzheng zhi* (Civil administration in local chronicles of Sichuan Province) (Chengdu: Sichuan renmin chubanshe, 1996), 216.

94. Yu, *Rongyu junren zhi zhiye zaizao,* 13.

95. Ibid.

4. Creating Gendered Images of the Soldier Figure in Literary Works

1. Wang Ke and Xu Sai, *Xiao Jun pingzhuan* (Biography of Xiao Jun) (Chongqing: Chongqing chubanshe, 1993), 8.

2. Ibid.

3. Xiao Jun, *Ren yu renjian: Xiao Jun huiyilu* (The person and his life: Memoirs of Xiao Jun) (Beijing: Zhongguo wenlian chubanshe, 2006), 25, 99, 191.

4. *Yangjia jiang* is a collection of folktales, plays, and novels detailing the exploits of a clan of generals (with the family name Yang) over four generations during China's Northern Song dynasty (960–1127). The stories recount the unflinching loyalty of the members of the Yang clan, and of how they defended the Song dynasty's borders from foreign invaders.

5. Xiao's essay "Nuo" was published in the literary supplement of *Shengjing shibao* (Shengjing times) in May 1929.

6. The Mukden Incident, or Manchurian Incident, was an event engineered by the Imperial Japanese Army on September 18, 1931, as a pretext for the Japanese invasion in 1931 of northeastern China, known as Manchuria.

7. Zhang Yuhong, "Ping *Bayuede xiangcun*" (A Review of *Village in August*), in *Zhongguo xiandai baibu zhongchangpian xiaoshuo lunxi, shang* (Reviews of a hundred novellas and novels in modern China, part 1), ed. Liu Zhongshu (Changchun: Jilin daxue chubanshe, 1986), 428.

8. Ibid.

9. Edgar Snow, "Introduction," in Xiao Jun, *Village in August,* trans. Tien Chün (New York: Smith and Durrell, 1942), xiii.

10. Xiao Jun, *Bayuede xiangcun* (Village in August) (Nanjing: Jiangsu wenyi chubanshe, 2010), 6.

11. Ibid.

12. Ibid., 26.

13. Ibid., 76.

14. Ibid., 7.

15. Ibid., 10.

16. Ibid.

17. Ibid., 23.

18. Ibid., 25.

19. Ibid., 92.

20. Ibid., 35.

21. Ibid., 15.

22. Ibid., 47.

23. Ibid., 36.

24. Ibid., 47.

25. Ibid.

26. Lu Xun, *Lu Xun quanji* (Collected works of Lu Xun), vol. 6 (Beijing: Renmin chubanshe, 2005), 296.

27. He Mu, "*Bayuede xiangcun*" (Village in August), *Shishi xinbao* (New newspapers of current affairs) (February 25, 1938). This source is reprinted from Zhang, "Ping *Bayuede xiangcun*," 429.

28. Leo Ou-fan Lee, "Literary Trends: The Road to Revolution," in *An Intellectual History of Modern China,* ed. Merle Goldman and Leo Ou-fan Lee (New York: Cambridge Univ. Press, 2002), 230.

29. Yi Qun, "Kangzhan yilai de Zhongguo baogao wenxue" (Chinese reportage literature since the breakout of the Anti-Japanese War), in *Zhongguo kangri zhanzheng shiqi dahoufang wenxue shuxi* (A compendium of literary works published in the interior during China's Anti-Japanese War), vol. 3, ed. Lin Mohan et al. (Chongqing: Chongqing renmin chubanshe, 1989), 1377–1378.

30. Yunzhong Shu, *Buglers on the Home Front: The Wartime Practice of the Qiyue School* (Albany: State Univ. of New York Press, 2000), 43.

31. The journal was named after the month of the Marco Polo Bridge Incident on July 7, 1937. The event was widely considered as the marker for the start of the Second Sino-Japanese War.

32. Shu, *Buglers on the Home Front,* 43.

33. Ibid., 44.

34. Ibid., 48. This original source of Hu Feng's direction on the reportage writing was in *Qiyue* (July) 2 (November 1937): 39.

35. Hu Feng, "Lun zhandouqi de yige zhandou de wenyi xingshi" (On a com-

bative literary genre in wartime), in *Hu Feng pinglunji* (Collections literary critics of Hu Feng), vol. 2 (Beijing: Renmin wenxue chubanshe, 1985), 16–24.

36. Shu, *Buglers on the Home Front*, 89.

37. Qiu Dongping, "Diqilian" (The seventh company) [1938], in *Qiu Dongping daibiaozuo* (Representative works by Qiu Dongping), ed. Zhongguo xiandai wenxueguan (Beijing: Huaxia chubanshe, 2011), 171. The English translation of this paragraph is drawn from Charles A. Laughlin, *Chinese Reportage: The Aesthetics of Historical Experience* (Durham, N.C.: Duke Univ. Press, 2002), 188.

38. Qiu, "Diqilian," 173. The translation of this paragraph is drawn from Shu, *Buglers on the Home Front*, 57.

39. Qiu, "Diqilian," 174.

40. Ibid. The translation of this paragraph is drawn from Laughlin, *Chinese Reportage*, 189.

41. Qiu, "Diqilian," 171.

42. Ibid., 175. The English translation is drawn from Laughlin, *Chinese Reportage*, 189.

43. Qiu, "Diqilian," 172.

44. Ibid., 174. The English translation is drawn from Laughlin, *Chinese Reportage*, 189.

45. Laughlin, *Chinese Reportage*, 190.

46. Qiu, "Diqilian," 175. The English translation is drawn from Laughlin, *Chinese Reportage*, 189.

47. Laughlin, *Chinese Reportage*, 190.

48. Qiu, "Diqilian," 175.

49. Shu, *Buglers on the Home Front*, 58.

50. Laughlin, *Chinese Reportage*, 191.

51. Ibid.

52. Qiu Dongping, "Honghuadi zhi shouyu" (The defense of Honghuadi) [1935], in *Qiu Dongping daibiaozuo* (Representative works by Qiu Dongping), ed. Zhongguo xiandai wenxueguan (Beijing: Huaxia chubanshe, 2011), 26.

53. Qiu Dongping, "Tongxunyuan" (Correspondent) [1932], in *Qiu Dongping daibiaozuo* (Representative works by Qiu Dongping), ed. Zhongguo xiandai wenxueguan (Beijing: Huaxia chubanshe, 2011), 32.

54. Ibid.

55. Ibid., 33.

56. Ibid., 34.

57. Qiu, "Diqilian," 175.

58. Qiu Dongping, "The Lieutenant Colonel" (Zhongxiao fuguan) [1938], in *Qiu Dongping daibiaozuo* (Representative works by Qiu Dongping), ed. Zhongguo xiandai wenxueguan (Beijing: Huaxia chubanshe, 2011), 46.

59. Ibid., 51.

60. Ibid., 52.

61. Ibid.

62. Shu, *Buglers on the Home Front,* 60.

63. Qiu Dongping, "Yige lianzhang de zhandou zaoyu" (A company commander's combat experience) [1938], in *Qiu Dongping daibiaozuo* (Representative works by Qiu Dongping), ed. Zhongguo xiandai wenxueguan (Beijing: Huaxia chubanshe, 2011), 206.

64. Ibid., 214.

65. Ibid.

66. Ibid.

67. The journal *Zhongguo junren* was initiated on February 20, 1925, in Guangzhou and ceased publication in March 1926.

68. Jin Huishu's birth and death years are unclear.

69. Li Zhilong, "Lujun junguan xuexiao zhaoshou nüsheng wenti" (The issue of recruiting woman cadets into the army officer academy), *Zhongguo junren* (Chinese soldiers) 6 (August 17, 1925): 47.

70. Many well-known female revolutionaries in modern Chinese history used to study at the Whampoa Military Academy, such as Zhao Yiman (1905–1936) and Hu Lanqi (1901–1994). After the Mukden Incident in 1937, the Communist soldier Zhao served as the political commissar of the 2nd Regiment of the Third Army of the Northeast Anti-Japanese United Army. She was captured and executed by Japanese forces in 1936. Hu organized and led a team of women who supported the Nationalist army's attempts to resist the Japanese invasion in 1937. She was appointed China's first female major general by the Nationalists' Central Military Commission. For Hu's military experiences, see Kristin Stapleton, "Hu Lanqi: Rebellious Woman, Revolutionary Soldier, Discarded Heroine, and Triumphant Survivor," in *The Human Tradition in Modern China,* ed. Kenneth Hammond and Kristin Stapleton (Lanham, Md.: Rowman and Littlefield, 2007), 166.

71. Ren Zipeng, "Dageming shidai Huangpu nübing" (Whampoa women cadets in the revolutionary era), *Wenhua Zhongguo* (Chinese culture) (July 14, 2012), http://cul.china.com.cn/lishi/2012-07/14/content_5161009.htm (accessed October 12, 2012).

72. Ibid.

73. The first English-language translation appeared in the United States as *Hsieh Pingying, Girl Rebel: The Autobiography of Hsieh Pingying, with Extracts from Her New War Diaries,* trans. Adet and Anor Lin (New York: John Day, 1940). An alternate translation appeared in England: *Hsieh Bing-ying, Autobiography of a Chinese Girl: A Genuine Autobiography,* trans. Tsui Chi (London: Allen and Unwin, 1943). Both translations were given a new lease on circulation through reprints, the first by Da Capo Press in 1975, the other by Pandora in 1986. For a new translation, see Xie Bingying, *A Woman Soldier's Own Story,* trans. Lily Chia Brissman and Barry Brissman (New York: Columbia Univ. Press, 2001). The information above is quoted from Patricia Sieber, "Introduction," in *Red Is Not the Only Color: Contemporary Chinese Fiction on Love and Sex between Women, Collected Stories,* ed. Patricia Sieber (Lanham, Md.: Rowman and Littlefield, 2001), 32.

74. Xie Bingying, "Yige nübingde zizhuan" (The autobiography of a female soldier), in *Xie Bingying daibiaozuo* (Representative works by Xie Bingying), ed. Zhongguo xiandai wenxueguan (Beijing: Huaxia chubanshe, 2009), 59.

75. Jing M. Wang, *When "I" Was Born: Women's Autobiography in Modern China* (Madison: Univ. of Wisconsin Press, 2008), 179.

76. Xie, "Yige nübingde zizhuan," 72.

77. Ibid., 88.

78. Ibid.

79. Ibid., 79.

80. Ibid., 84.

81. Xie Bingying, "Congjun riji" (Army diaries), in *Xie Bingying daibiaozuo* (Representative works by Xie Bingying), ed. Zhongguo xiandai wenxueguan (Beijing: Huaxia chubanshe, 2009), 5.

82. Ibid., 8.

83. Ibid., 39.

84. Xie Bingying, "Congjun riji," 6.

85. Yi Ming's birth and death years are unclear.

86. Yi Ming, "Zai zhandi fuwu de Xie Bingying" (Xie Bingying in the battlefield service), in *Kangzhan zhongde nüzhanshi* (Women soldiers during the Anti-Japanese War), ed. Shen Zijiu (N.p.: Zhanshi chubanshe, [1938]), 42.

5. The Construction of the Soldier Ideal by Educated Youths

1. Jay Taylor, *The Generalissimo's Son: Chiang Ching-kuo and the Revolutions in China and Taiwan* (Cambridge, Mass.: Harvard Univ. Press, 2000), 119.

2. Sidney Chang and Ramon Hawley Myers, eds., *The Storm Clouds Clear over China: The Memoir of Chen Lifu, 1900–1993* (Stanford, Calif.: Hoover Institution Press, 1994), 168.

3. Mitsutoshi Hanyu, "From Campus to Battle: Student Mobilization and Transition of Japanese Imperial Military's Draft Policies," *Military History Research Annual Report,* vol. 12 (March 2009): 108–114, http://www.nids.mod.go.jp/publication/senshi/pdf/200903/08.pdf (accessed September 21, 2016).

4. Ibid.

5. Chiang Kai-shek, Chiang Kai-shek diaries, March 11, 1943, box 43, folder 2, Hoover Institution Archives.

6. Ibid., July 11, 1937, box 39, folder 13.

7. Ibid., December 24, 1935, box 38, folder 8.

8. Bai Chongxi, *Bai Chongxi huiyilu* (Memoirs of Bai Chongxi) (Beijing: Jiefangjun chubanshe, 1987), 218.

9. Guo Shaoyi, *Qingnian yuanzhengjun zhilüe* (The history of the Educated Youth Expeditionary Army) (Taibei: Youshi wenhua shiye gongsi, 1987), 176.

10. Hans van de Ven, *War and Nationalism in China, 1925–1945* (New York: Routledge Curzon, 2003), 17.

11. Ibid., 276.

12. Taylor, *The Generalissimo's Son,* 118.

13. Michael Dillon, *China: A Modern History* (New York: I. B. Tauris, 2010), 244.

14. Peng Xunhou, *Shijie fanfaxisi zhanzheng zhong de Zhongguo* (China in the World Anti-Fascism War) (Beijing: Wuzhou chuanbo chubanshe, 2005), 109.

15. Kenneth Pletcher, "The Late Republican Period and the War against Japan," in *The History of China,* ed. Kenneth Pletcher (New York: Britannica Educational Publishing, 2011), 287.

16. Taylor, *The Generalissimo's Son,* 119.

17. Jiang Pei, "Zhanshi zhishi qingnian congjun yundong shuping" (A study on the Campaign to Mobilize Educated Youths to Join the Army during the Anti-Japanese War), *Kangri zhanzheng yanjiu* (Studies on the Anti-Japanese War) 1 (2004): 67.

18. Ibid., 68.

19. Ibid.

20. He Yingqin, *Banian kangzhan zhi jingguo* (The process of the eight-year resistance war against the Japanese) (Taibei: Liming wenhua shiye gongsi, 1982), 233.

21. Guo, *Qingnian yuanzhengjun zhilüe,* 12.

22. Chiang Kai-shek, "Gao zhishi qingnian congjun shu" (A speech to call on educated youths to join the army), in Guo Shaoyi, *Qingnian yuanzhengjun zhilüe* (The history of the Educated Youth Expeditionary Army) (Taibei: Youshi wenhua shiye gongsi, 1987), 201.

23. Ibid., 203.

24. Ibid., 201.

25. Ibid., 207.

26. Ibid., 201.

27. Ibid.

28. Chiang Kai-shek, "Duiyu zhishi qingnian congjun yundong zhi zhishi" (The direction on the campaign of enlisting educated youths), in Guo Shaoyi, *Qingnian yuanzhengjun zhilüe* (The history of the Educated Youth Expeditionary Army) (Taibei: Youshi wenhua shiye gongsi, 1987), 190.

29. Ibid., 187.

30. Chang Rui-te, "The National Army from Whampoa to 1949," in *A Military History of China,* ed. David Graff and Robin Higham (Lexington: Univ. Press of Kentucky, 2012), 204.

31. Chiang Kai-shek, "Qingnian yuanzhengjun bianxunde yaozhi" (The key principles in training the Educated Youth Expeditionary Army) [1944], in Guo Shaoyi, *Qingnian yuanzhengjun zhilüe* (The history of the Educated Youth Expeditionary Army) (Taibei: Youshi wenhua shiye gongsi, 1987), 210.

32. Ibid.

33. Ibid.

34. Chiang, "Gao zhishi qingnian congjun shu," 207.

35. Chiang Kai-shek, "Dui congjun xuesheng xunhua" (A lecture to student soldiers), in Guo Shaoyi, *Qingnian yuanzhengjun zhilüe* (The history of the Educated Youth Expeditionary Army) (Taibei: Youshi wenhua shiye gongsi, 1987), 177.

36. Chiang, "Duiyu zhishi qingnian congjun yundong zhi zhishi," 186.

37. Chiang, "Gao zhishi qingnian congjun shu," 204.

38. Ibid.

39. Chiang Kai-shek, "Qingnian yuanzhengjun bianlian zhi tezhi yu jiaoyu yaoxiang" (The peculiarity and key points in training the Educated Youth Expeditionary Army), in Guo Shaoyi, *Qingnian yuanzhengjun zhilüe* (The history of the Educated Youth Expeditionary Army) (Taibei: Youshi wenhua shiye gongsi, 1987), 238.

40. Ibid.

41. Ibid.

42. Chiang, "Dui congjun xuesheng xunhua," 182.

43. Ibid.

44. Ibid.

45. Chiang Kai-shek, "Qingnian yuanzhengjun bianlian guanxun de fangzhen yu yaoling" (The principles and main points in training the Educated Youth Expeditionary Army), in *Xian zongtong Jiang gong sixiang yanlun zongji* (Comprehensive collections of the thoughts and speeches of deceased President Chiang), vol. 20, ed. Qin Xiaoyi (Taibei: Zhongyang dangshi weiyuanhui, 1984), 565.

46. Sun Yuqin, "Kangzhan moqide shiwan zhishi qingnian congjun yundong shuping" (A review on the Campaign to Mobilize Educated Youths to Join the Army in the last stage of the Anti-Japanese War), *Kangri zhanzheng yanjiu* (Studies on the Anti-Japanese War) 3 (2010): 25.

47. Chiang, "Qingnian yuanzhengjun bianlian zhi tezhi yu jiaoyu yaoxiang," 243.

48. Chiang, "Qingnian yuanzhengjun bianxunde yaozhi," 215.

49. Ibid.

50. Taylor, *The Generalissimo's Son,* 119.

51. Zhang Jingcang's birth and death years are unclear.

52. Zhang Jingcang, "Xinzhanshi xinzuofeng" (New soldier and new styles), in *Huoyue de qingnianjun* (The active Youth Army), ed. Luo Shiyang (N.p.: Qingnian chubanshe, 1946), 92.

53. The writings by the Youth Army soldiers published in the *Zhongyang ribao* were collected in several volumes. The writings analyzed in this chapter are drawn from the following three volumes: Junshi weiyuanhui quanguo zhishi qingnian zhiyuan congjun bianlian zongjianbu, ed., *Qingnian yuanzhengjun chuangzuo zuopinxuan* (Selections of the works by the Educated Youth Expeditionary Army soldiers) (Chongqing: Junshi weiyuanhui quanguo zhishi qingnian zhiyuan congjun bianlian zongjianbu, 1945); Junshi weiyuanhui quanguo zhishi qingnian zhiyuan congjun bianlian zongjianbu, ed., *Qingnian yuanzhengjun chuangzuo*

zuopinxuan xuji (Continued selections of the works by the Educated Youth Expeditionary Army soldiers) (Chongqing: Junshi weiyuanhui quanguo zhishi qingnian zhiyuan congjun bianlian zongjianbu, 1945); and Shiyang, ed., *Huoyue de qingnianjun.*

54. Zhang Qing, "Junzhong wenxue" (Literature and the army), in *Qingnian yuanzhengjun chuangzuo zuopinxuan* (Selections of the works by the Educated Youth Expeditionary Army soldiers), ed. Junshi weiyuanhui quanguo zhishi qingnian zhiyuan congjun bianlian zongjianbu (Chongqing: Junshi weiyuanhui quanguo zhishi qingnian zhiyuan congjun bianlian zongjianbu, 1945), 66.

55. Hou Bingchen, "Xin jun xin shenghuo" (New army and new life), in *Huoyue de qingnianjun* (The active Youth Army), ed. Luo Shiyang (N.p.: Qingnian chubanshe, 1946), 84.

56. Zha Mi, "Laodong yu yule" (Labor and Entertainment), in *Huoyue de qingnianjun* (The active Youth Army), ed. Luo Shiyang (N.p.: Qingnian chubanshe, 1946), 24.

57. Lian Lulu, "Wanhui" (An evening party), in *Qingnian yuanzhengjun chuangzuo zuopinxuan* (Selections of the works by the Educated Youth Expeditionary Army soldiers), ed. Junshi weiyuanhui quanguo zhishi qingnian zhiyuan congjun bianlian zongjianbu (Chongqing: Junshi weiyuanhui quanguo zhishi qingnian zhiyuan congjun bianlian zongjianbu, 1945), 18.

58. Zhao Tingjun, "Jiyu fulao" (To the elders in the family), in *Qingnian yuanzhengjun chuangzuo zuopinxuan* (Selections of the works by the Educated Youth Expeditionary Army soldiers), ed. Junshi weiyuanhui quanguo zhishi qingnian zhiyuan congjun bianlian zongjianbu (Chongqing: Junshi weiyuanhui quanguo zhishi qingnian zhiyuan congjun bianlian zongjianbu, 1945), 29.

59. Luo Qiqian, "Qingnianjun yu qingnian Zhongguo" (The Youth Army and young China), in *Qingnian yuanzhengjun chuangzuo zuopinxuan xuji* (Continued selections of the works by the Educated Youth Expeditionary Army soldiers), ed. Junshi weiyuanhui quanguo zhishi qingnian zhiyuan congjun bianlian zongjianbu (Chongqing: Junshi weiyuanhui quanguo zhishi qingnian zhiyuan congjun bianlian zongjianbu, 1945), 4.

60. Jia Yunfu, "Qingnian lushang" (On the road being built by the youth), in *Qingnian yuanzhengjun chuangzuo zuopinxuan* (Selections of the works by the Educated Youth Expeditionary Army soldiers), ed. Junshi weiyuanhui quanguo zhishi qingnian zhiyuan congjun bianlian zongjianbu (Chongqing: Junshi weiyuanhui quanguo zhishi qingnian zhiyuan congjun bianlian zongjianbu, 1945), 4.

61. Zha, "Laodong yu yule," 23.

62. Zhao, "Jiyu fulao," 28.

63. Zheng Bingsen, "Gei muqinmen" (To the mothers), in *Qingnian yuanzhengjun chuangzuo zuopinxuan* (Selections of the works by the Educated Youth Expeditionary Army soldiers), ed. Junshi weiyuanhui quanguo zhishi qingnian zhiyuan congjun bianlian zongjianbu (Chongqing: Junshi weiyuanhui quanguo zhishi qingnian zhiyuan congjun bianlian zongjianbu, 1945), 33.

64. Stacey Bieler, *"Patriots" or "Traitors"?: A History of American-Educated Chinese Students* (New York: Routledge, 2009), 171.

65. Chen Lang Tung, "Christian Education in China," *Chinese Students' Monthly* 18, no. 3 (1923): 21.

66. Xiaoping Cong, *Teachers' Schools and the Making of the Modern Chinese Nation-State, 1897–1937* (Vancouver, Canada: Univ. of British Columbia Press, 2007), 83.

67. Susan Rigdon, "National Salvation: Teaching Civic Duty in China's Christian Colleges," in *China's Christian Colleges: Cross-Cultural Connections, 1900–1950*, ed. Daniel Bays and Ellen Widmer (Stanford, Calif.: Stanford Univ. Press, 2009), 201.

68. Kristen Mulready-Stone, *Mobilizing Shanghai Youth: CCP Internationalism, GMD Nationalism and Japanese Collaboration* (New York: Routledge, 2015).

69. Zhao, "Jiyu fulao," 29.

70. Qiu Hongyi, "Women zai Hufeng" (We are in Hufeng), in *Huoyue de qingnianjun* (The active Youth Army), ed. Luo Shiyang (N.p.: Qingnian chubanshe, 1946), 87–88.

71. Chen Can, "Womende xin feidao zhanchang" (Our hearts have flown to the battlefield), in *Huoyue de qingnianjun* (The active Youth Army), ed. Luo Shiyang (N.p.: Qingnian chubanshe, 1946), 96.

72. Ouyang Wenhui, "Zhenggongban shenghuo zhuishu" (Memories of the life in the class for political workers), in *Huoyue de qingnianjun* (The active Youth Army), ed. Luo Shiyang (N.p.: Qingnian chubanshe, 1946), 24.

73. Zhang Jingcang, "Xinzhanshi xinzuofeng" (New soldiers and new styles), in *Huoyue de qingnianjun* (The active Youth Army), ed. Luo Shiyang (N.p.: Qingnian chubanshe, 1946), 94.

74. Ibid.

75. Shen Yi, "Kaishile junying shenghuo" (The army life has started), in *Huoyue de qingnianjun* (The active Youth Army), ed. Luo Shiyang (N.p.: Qingnian chubanshe, 1946), 72.

76. Zhang Zhengquan, "Junshiduide yingzhong shenghuo" (The army life in the base), in *Huoyue de qingnianjun* (The active Youth Army), ed. Luo Shiyang (N.p.: Qingnian chubanshe, 1946), 76.

77. Ouyang, "Zhenggongban shenghuo zhuishu," 25.

78. Hou, "Xin jun xin shenghuo," 82.

79. Ibid., 83.

80. "Chufa xingjun ruying" (Marching to the military base), in *Huoyue de qingnianjun* (The active Youth Army), ed. Luo Shiyang (N.p.: Qingnian chubanshe, 1946), 41.

81. Shen, "Kaishile junying shenghuo," 72.

82. "Chufa xingjun ruying," 38.

83. Ibid., 40.

6. The Army-People Bond in Mass Culture in Wartime Yan'an

1. Throughout the Second Sino-Japanese War of 1937 to 1945, the CCP formed border region (*bianqu*) governments in north and central China. In September 1937, the Communists formally announced the formation of the Shaan-Gan-Ning border region government. Nominally subordinate to the GMD government, the border region in fact subjected itself only to the control of the CCP and implemented only CCP policies.

2. John Gittings, *The Role of the Chinese Army* (London: Oxford Univ. Press, 1967), 303.

3. William Wei, "Mao and the Red Army," in *A Military History of China*, ed. David A. Graff and Robin Higham (Lexington: Univ. Press of Kentucky, 2012), 235.

4. Evans Fordyce Carlson, *The Chinese Army, Its Organization and Military Efficiency* (New York: Institute of Pacific Relations, 1940), 36.

5. Lei Zhihua and Li Zhongquan, eds., *Shaan-Gan-Ning bianqu minzheng gongzuo ziliao xuanbian* (Selective materials on civil administration work in Shaan-Gan-Ning border region) (Xi'an: Shaanxi renmin chubanshe, 1992), 209.

6. Han Yanlong and Chang Zhaoru, eds., *Zhongguo xinminzhuzhuyi geming shiqi genjudi fazhi wenxian xuanbian, diyijuan* (Selections of legislative sources in the revolutionary bases during China's new democratic revolution), vol. 1 (Beijing: Zhongguo shehui kexue chubanshe, 1981).

7. The Eighth Route Army and the New Fourth Army were the CCP's two main military forces active during the Second Sino-Japanese War (1937–1945). Nominally, they were units of the National Revolutionary Army of the Republic of China led by Chinese Nationalists.

8. Mao Zedong, "Talks at the Yenan Forum on Literature and Art," in Mao Zedong, *Selected Works of Mao Zedong*, vol. 3 (Beijing: Foreign Languages Press, 1967), 73.

9. David Holm, *Art and Ideology in Revolutionary China* (Oxford: Clarendon Press, 1991), 93.

10. Mao, "Talks at the Yenan Forum on Literature and Art," 73.

11. Ibid., 74.

12. Ibid.

13. Ibid.

14. Zhou Yang, "Makesi zhuyi yu wenyi" (Marxism and literatures and arts), 1944, in *Wenyi lilun* (Theories on literature and arts), ed. Yan'an wenyi congshu bianweihui, Yan'an wenyi congshu (Series of Yan'an literatures and arts), vol. 1 (Changsha: Hunan renmin chubanshe, 1983), 215.

15. Mark Selden, *China in Revolution: The Yenan Way Revisited* (New York: M. E. Sharpe, 1995), 187.

16. Ibid., 197.

17. Ibid.

18. *Kangri zhanzheng shiqi jiefangqu gaikuang* (General situation in the liberation areas during the Anti-Japanese War) (Beijing: Renmin chubanshe, 1953), 15.

19. Selden, *China in Revolution,* 197. As the second section of this paper will show, the slogan of lessening the burden of the military on peasants was repeatedly articulated in the *yangge* plays. According to Selden, the army played a leading role in the production movement, and its activities were representative of those of other groups. The model unit was Wang Zhen's (1908–1993) 359th Brigade, which since 1938 had been the leader in military production at Nan-ni-wan.

20. Shaanxisheng dang'anguan and Shaanxisheng shehui kexueyuan, eds., *Shaan-Gan-Ning bianqu wenjian xuanbian* (Selection of documents on the Shaanxi-Gansu-Ningxia Border Region), vol. 7 (Beijing: Dang'an chubanshe, 1988), 12.

21. Ibid.

22. Holm, *Art and Ideology in Revolutionary China,* 9.

23. *Kangri zhanzheng shiqi jiefangqu gaikuang,* 18.

24. Wang Shoudao, "Chudong zhong de hongjun" (The Red Army in motion), 1937, in *Baogaowenxue* (documentary literatures), ed. Huang Gang, Zhongguo jiefangqu wenxue shuxi (Series of literary works in the Chinese liberation areas) (Chongqing: Chongqing chubanshe, 1992), 591.

25. *Kangri zhanzheng shiqi jiefangqu gaikuang,* 19.

26. Wang Shoudao, "Chudong zhong de hongjun," 591.

27. Dwarkanath Kotnis, "The most important time in my life," 1939, trans. Liu Chuangyuan, in *Waiguo renshi zuopin* (Selections of writings by foreigners), ed. Israel Epstein and Gao Liang, Zhongguo jiefangqu wenxue shuxi (Series of literary works in the Chinese liberation areas) (Chongqing: Chongqing chubanshe, 1992), 751.

28. "Yan'an yanchu jumu" (*Yangge* dramas performed in Yan'an), 1945, in *Kangri zhanzheng shiqi Yan'an ji gekangri minzhu genjudi wenxue yundong ziliao* (Materials on the literary movements in Yan'an and other anti-Japanese democratic base areas), ed. Liu Zengjie et al. (Taiyuan: Shanxi renmin chubanshe, 1983), 333.

29. Holm, *Art and Ideology in Revolutionary China,* 251.

30. Ibid., 130.

31. Zhao Chaogou, *Yan'an yiyue* (One month in Yan'an) (Washington D.C.: Association of Research Libraries, 1969), 103.

32. Zhongguo shehui kexueyuan xinwen yanjiusuo zhongguo baokanshi yanjiushi, ed., *Yan'an wencui* (Literature collections in Yan'an) (Beijing: Beijing chubanshe, 1984), 508.

33. Xue Xiaoxu and Du Lei, "Yan'an shiqi de xin yangge yundong jiedu" (Interpretation of the new *yangge* movement in Yan'an), *Wenshi ziliao* (Materials on literature and history) 33 (November 2008): 78.

34. Holm, *Art and Ideology in Revolutionary China*, 271.

35. Zhou Yang, "Biaoxian xin de qunzhong de shidai" (The era of writing about new mass), 1944, in *Wenxue yundong lilun* (Theories on literary movements), ed. Hu Cai, Zhongguo jiefangqu wenxue shuxi (Series of literary works in the Chinese liberation areas) (Chongqing: Chongqing chubanshe, 1992), 534.

36. John S. Service, *Lost Chance in China*, trans. Luo Qing and Zhao Zhongqiang, in *Waiguo renshi zuopin* (Selections of writings by foreigners), ed. Israel Epstein and Gao Liang, Zhongguo jiefangqu wenxue shuxi (Series of literary works in the Chinese liberation areas) (Chongqing: Chongqing chubanshe, 1992), 1248.

37. Zhou Yang, "Biaoxian xin de qunzhong de shidai," 541.

38. Ibid.

39. Ma Jianxiang, *Shi'er ba liandao* (Twelve sickles), 1938, in *Yanggeju* (The *yangge* drama), ed. Yan'an wenyi congshu bianweihui, Yan'an wenyi congshu (Series of Yan'an literatures and arts), vol. 7 (Changsha: Hunan renmin chubanshe, 1983).

40. Lianfangjun zhengzhibu xuanchuandui, *Zhang Zhiguo* (The soldier Zhang Zhiguo), 1943, in *Yanggeju* (The *yangge* drama), ed. Yan'an wenyi congshu bianweihui, Yan'an wenyi congshu (Series of Yan'an literatures and arts), vol. 7 (Changsha: Hunan renmin chubanshe, 1983), 68–94.

41. Lianfangjun zhengzhibu xuanchuandui, *Jun aimin, min yongjun* (The army loves the people, and the people support the army), 1943, in *Yanggeju* (The *yangge* drama), ed. Yan'an wenyi congshu bianweihui, Yan'an wenyi congshu (Series of Yan'an literatures and arts), vol. 7 (Changsha: Hunan renmin chubanshe, 1983), 237–258.

42. Lianfangjun zhengzhibu xuanchuandui, *Liu Shunqing* (Soldier Liu Shunqing), 1944, in *Yanggeju* (The *yangge* drama), ed. Yan'an wenyi congshu bianweihui, Yan'an wenyi congshu (Series of Yan'an literatures and arts), vol. 7 (Changsha: Hunan renmin chubanshe, 1983), 281–324.

43. Lianfangjun zhengzhibu xuanchuandui, *Niu Yonggui guacai* ((The injured soldier Niu Yonggui), 1944, in *Yanggeju* (The *yangge* drama), ed. Yan'an wenyi congshu bianweihui, Yan'an wenyi congshu (Series of Yan'an literatures and arts), vol. 7 (Changsha: Hunan renmin chubanshe, 1983), 68–94.

44. Jin Dongping, *Yan'an jianwenlu* (Documents of what I saw and heard in Yan'an) (Washington, D.C.: Association of Research Libraries, 1969), 96.

45. Israel Epstein, "Xinminzhu zhuyi houfang jianwen" (Observation of new democracy in the border region), 1947, trans. Chen Yaohua and Xie Nianfei, in *Waiguo renshi zuopin* (Selections of writings by foreigners), ed. Israel Epstein and Gao Liang, Zhongguo jiefangqu wenxue shuxi (Series of literary works in the Chinese liberation areas) (Chongqing: Chongqing chubanshe, 1992), 925.

46. Sulamith Heins Potter, "Cultural Construction of Emotion in Rural Chinese Social Life," *Ethos* 16, no. 2 (June 1988): 208.

47. Feng Mu, Yang Sizhong, and Huang Gang, "Women de budui zai shanlin

li" (Our army in the mountains and forests), in *Baogaowenxue* (Documentary literatures), ed. Yan'an wenyi congshu bianweihui, Yan'an wenyi congshu (Series of Yan'an literatures and arts), vol. 6 (Changsha: Hunan renmin chubanshe, 1983), 183.

Bibliography

Archives

Chiang Kai-shek Diaries Hoover Institution Archives.

Records of YMCA international work in China (Y.USA.9-2-4). University of Minnesota Libraries, Kautz Family YMCA Archives.

Chinese-Language Primary Sources

Bai, Chongxi. *Bai Chongxi huiyilu* (Memoirs of Bai Chongxi). Beijing: Jiefangjun chubanshe, 1987.

Chen, Junming. *Yige jinjide huyu: shangbing gongzuo zhi lilun, shiji yu jianyi* (An emergent call: The theory, practice, and suggestion for wounded soldier work). Changsha: Zhongxin gongsi, 1939.

Cheng, Tingrong. "Chengdu zhongyang junxiao xuechao jishi" (The cadet uprising in the Chengdu Branch Campus of the Central Army Officer Academy) [1963]. In *Guomindang zhongyang lujun xuexiao yu junshi zhuanke xuexiao* (The Nationalist Party's Central Army Officer Academy and colleges of military specialties), edited by Wen Wen, 56–58. Beijing: Zhongguo wenshi chubanshe, 2010.

Chiang, Ching-kuo. "Gao fushang jiangshi shu—wei qingzhu sanshinian yuandan er zuo" (To the wounded soldiers—a speech written for the celebration of the new year of 1941). *Shangbing zhi you* (Friends of the wounded soldiers) 9 (January 1, 1941): 8–10.

Deng, Wenyi, ed. *Huangpu xunlianji* (Selections of the training materials at the Whampoa Military Academy). Wuhan: Guofangbu xinwenju, 1938.

Du, Congrong. *Huangpu junxiao zhi chuangjian ji dongzheng beifa zhi huiyi* (Memories of the foundation of the Whampoa Military Academy and the Eastern and Northern Expeditions). Taibei: Shunren caise yinzhi youxian gongsi, 1975.

Epstein, Israel, and Gao Liang, eds. *Waiguo renshi zuopin* (Selections of writings by foreigners). Zhongguo jiefangqu wenxue shuxi (Series of literary works in the Chinese liberation areas). Chongqing: Chongqing chubanshe, 1992.

Fu, Sinian. "Duan Shengwu xiansheng zhuan" (Biography of Mr. Duan Shengwu). *Wenshi zazhi* (Journal of literature and history) 5, no. 7 (1945): 47.

Guomin zhengfu junshi weiyuanhui zhengzhibu, ed. *Huangpu xunlianji xuanji* (Selections from the collection of the training materials at the Whampoa Military Academy). N.p.: Guomin zhengfu junshi weiyuanhui zhengzhibu, 1938.

Guomindang zhongyang lujun junguan xuexiao xiaowu, ed. *Zhongyang lujun junguan xuexiao shigao* (History of the Central Army Officer Academy), vol. 7. Nanjing: Guomindang zhongyang lujun junguan xuexiao xiaowu, 1936.

Han, Yanlong, and Chang Zhaoru, eds. *Zhongguo xinminzhuzhuyi geming shiqi genjudi fazhi wenxian xuanbian, diyijuan* (Selections of legislative sources in the revolutionary bases during China's new democratic revolution), vol. 1. Beijing: Zhongguo shehui kexue chubanshe, 1981.

He, Yingqin. *Banian kangzhan zhi jingguo* (The process of the eight-year resistance war against the Japanese). Taibei: Liming wenhua shiye gongsi, 1982.

Hu, Cai, ed. *Wenxue yundong lilun* (Theories on literary movements). Zhongguo jiefangqu wenxue shuxi (Series of literary works in the Chinese liberation areas). Chongqing: Chongqing chubanshe, 1992.

Hu, Feng. *Hu Feng pinglunji* (Collected literary critics of Hu Feng), vol. 2. Beijing: Renmin wenxue chubanshe, 1985.

Hu, Lanqi. *Zhandi ernian* (Two years at the front line). Ji'an: Laodong funü zhandi fuwutuan, 1939.

Huang, Gang, ed. *Baogaowenxue* (Documentary literatures). Zhongguo jiefangqu wenxue shuxi (Series of literary works in the Chinese liberation areas). Chongqing: Chongqing chubanshe, 1992.

Huangpu xunlianji diyiji jingshen xunlian (Volume one on civic education in the training materials at the Whampoa Military Academy). N.p.: N.p., [1925].

Jin, Dongping. *Yan'an jianwenlu* (Documents of what I saw and heard in Yan'an). Washington, D.C.: Association of Research Libraries, 1969.

Jin, Mingsheng. *Zhonghua minguo xianfa cao'an shiyi* (Illustration on the Republic of China Constitution draft). Shanghai: Shijie shuju, 1936.

Junshi weiyuanhui junxunbu. *Zhanshi lujun jiaoyuling cao'an* (The draft of education decree in wartime period). Chongqing: Junshi weiyuanhui junxunbu, 1944.

Junshi weiyuanhui quanguo zhishi qingnian zhiyuan congjun bianlian zongjianbu, ed. *Qingnian yuanzhengjun chuangzuo zuopinxuan* (Selections of the works by the Educated Youth Expeditionary Army soldiers). Chongqing: Junshi weiyuanhui quanguo zhishi qingnian zhiyuan congjun bianlian zongjianbu, 1945.

———. *Qingnian yuanzhengjun chuangzuo zuopinxuan xuji* (Continued selections of the works by the Educated Youth Expeditionary Army soldiers). Chongqing: Junshi weiyuanhui quanguo zhishi qingnian zhiyuan congjun bianlian zongjianbu, 1945.

Junshi weiyuanhui weiyuanzhang Nanchang xingying, ed. *Shibing shizi keben, disance* (Literacy textbooks for soldiers, level 3). Shanghai: Zhonghua shuju, 1935.

Junzhengbu bingyishu yizhengsi xuanchuanbu, ed. *Bingyi xuanchuan ji youdai zhengshu faling huibian* (Compilation of conscription propaganda as well as laws and regulations on favorably treating soldiers' dependents). Chongqing: Junzhengbu bingyishu yizhengsi xuanchuanbu, 1943.

Kangri zhanzheng shiqi jiefangqu gaikuang (General situation in the liberation areas during the Anti-Japanese War). Beijing: Renmin chubanshe, 1953.

Lan, Hai. *Zhongguo kangzhan wenyi shi* (A history of the literatures in China's Anti-Japanese War). Jinan: Shandong wenyi chubanshe, 1984.

Lei, Zhihua, and Li Zhongquan, eds. *Shaan-Gan-Ning bianqu minzheng gongzuo ziliao xuanbian* (Selective materials on civil administration work in Shaan-Gan-Ning border region). Xi'an: Shaanxi renmin chubanshe, 1992.

Li, Zhilong. "Lujun junguan xuexiao zhaoshou nüsheng wenti" (The issue of recruiting woman cadets into the army officer academy). *Zhongguo junren* (Chinese soldiers) 6 (1925): 44–48.

Liang, Xiaochu. *Zhonghua jidujiao qingnianhui zhanqu fuwu quanguo weiyuanhui baogaoshu* (Reports of the national committee for war area service led by the Chinese YMCA). N.p.: Zhonghua jidujiao qingnianhui, 1933.

Lin, Mohan, et al., eds. *Zhongguo kangri zhanzheng shiqi dahoufang wenxue shuxi* (A compendium of literary works published in the interior during China's Anti-Japanese War), vol. 3. Chongqing: Chongqing renmin chubanshe, 1989.

Liu, Liangmo. "Zhanshide junren fuwu" (Soldier service in wartime). Hankou: Xinzhi shudian, 1938.

Liu, Naifu. *Zhandi fuwu gongzuo yu jingyan* (Experiences in performing war area service work). Hankou: Shenghuo shudian, 1938.

Liu, Zengjie, et al., eds. *Kangri zhanzheng shiqi Yan'an ji gekangri minzhu genjudi wenxue yundong ziliao* (Materials on the literary movements in Yan'an and other anti-Japanese democratic base areas). Taiyuan: Shanxi renmin chubanshe, 1983.

Lu, Xun. *Lu Xun quanji* (Collected works of Lu Xun), vol. 6. Beijing: Renmin chubanshe, 2005.

Luo, Derong. *Xinbian junren jingshen jiaoyu* (New edition of civic education for soldiers). N.p.: Junweihui tewutuan zhengxunchu xianbing jiaodao zongdui zhengxunchu, 1932.

Luo, Shiyang, ed. *Huoyue de qingnianjun* (The active Youth Army). N.p.: Qingnian chubanshe, 1946.

Mao, Zedong. *Selected Works of Mao Zedong*, vol. 3. Beijing: Foreign Languages Press, 1967.

Qi, Xiangming. "Jiuyiba qianhoude zhongyang junxiao" (The Central Army Officer Academy around the period of the Mukden Incident on September 18, 1931) [1963]. In *Guomindang zhongyang lujun xuexiao yu junshi zhuanke xuexiao* (The Nationalist Party's Central Army Officer Academy and colleges of military specialties), edited by Wen Wen, 17–19. Beijing: Zhongguo wenshi chubanshe, 2010.

Qian, Daquan. "Zhongyang junxiao xiqian jishi" (Narratives on the relocation of the Central Army Officer Academy to the west) [1963]. In *Guomindang zhongyang lujun xuexiao yu junshi zhuanke xuexiao* (The Nationalist Party's Central Army Officer Academy and colleges of military specialties), edited by Wen Wen, 37–40. Beijing: Zhongguo wenshi chubanshe, 2010.

Qin, Feng, ed. *International News Photographs in China during the Second Sino-Japanese War* (Kangzhan Zhongguo guoji tongxun zhaopian). Guilin: Guangxi shifan daxue chubanshe, 2008.

Qin, Xiaoyi, ed. *Xian zongtong Jiang gong sixiang yanlun zongji* (Comprehensive collections of the thoughts and speeches of deceased President Chiang), vols. 20 and 25. Taibei: Zhongyang dangshi weiyuanhui, 1984.

Quanguo jidujiao qingnianhui junren fuwubu tonggong, ed. *Kangri jiuwang shiride lishi huigu* (The historical review of the years during the Anti-Japanese War). Ha'erbin: Shuangchengshi yinchuachang, 1994.

Rongyu junren zhiye xiedaohui, ed. *Rongyu junren zhiye xiedaohui diyi niandu gongzuo baogao* (Reports on the first-year work of the Vocational Coordination Association for the Disabled Veterans). Chongqing: Rongyu junren zhiye xiedaohui, 1941.

Shaanxisheng dang'anguan, and Shaanxisheng shehui kexueyuan, eds. *Shaan-Gan-Ning bianqu wenjian xuanbian* (Selections of documents on the Shaanxi-Gansu-Ningxia Border Region). Beijing: Dang'an chubanshe, 1988.

Shen, Jianzhong, ed. *Cartoons on the Second Sino-Japanese War* (Kangzhan manhua). Shanghai: Shanghai shehui kexueyuan chubanshe, 2005.

Shen, Pu, ed. *Fan Changjiang xinwen wenji* (Collections of Fan Changjiang's news reports). Beijing: Xinhua chubanshe, 2001.

Shen, Zhenchuan. "Wuhan fenxiao dibaqi xuesheng kangri shiwei shijian qianhou" (A demonstration against the Japanese by the cadets in the eighth class of the Wuhan Branch Campus of the Central Army Officer Academy) [1964]. In *Guomindang zhongyang lujun xuexiao yu junshi zhuanke xuexiao* (The Nationalist Party's Central Army Officer Academy and colleges of military specialties), edited by Wen Wen, 59–62. Beijing: Zhongguo wenshi chubanshe, 2010.

Shen, Zijiu, ed. *Kangzhan zhongde nüzhanshi* (Women soldiers during the Anti-Japanese War). N.p.: Zhanshi chubanshe, [1938].

Shi, Mei, et al., eds. *Kangzhan jianguo gangling wenda* (Questions and answers on the guidelines of resisting the Japanese and building the state). Shanghai: Shenghuo shudian, 1938.

Sichuansheng difangzhi bianzuan weiyuanhui. *Sichuansheng zhi, minzheng zhi* (Civil administration in local chronicles of Sichuan Province). Chengdu: Sichuan renmin chubanshe, 1996.

Sun, Tzu. *Sunzi bingfa* (The art of war). Translated by Lionel Giles. El Paso, Tex.: El Paso Norte Press, 2005.

Tan, Dingyuan. "Zhongyang junxiao shi'erqi jianwen" (Experiences in the twelfth class of the Central Army Officer Academy) [1961]. In *Guomindang zhongyang lujun xuexiao yu junshi zhuanke xuexiao* (The Nationalist Party's Central Army Officer Academy and colleges of military specialties), edited by Wen Wen, 20–23. Beijing: Zhongguo wenshi chubanshe, 2010.

Tang, Degang. *Li Zongren huiyi lu* (Memoirs of Li Zongren) [1958]. Hongkong: Nanyue chubanshe, 1986.

Tian, Han, ed. *Zhandi guilai* (Return from the battlefield). N.p.: Zhanshi chubanshe, [1938].

Wang, Zhuochao. "Yi Nanjing zhongyang junxiao" (Remembering the Central Army Officer Academy at Nanjing) [1982]. In *Guomindang zhongyang lujun xuexiao yu junshi zhuanke xuexiao* (The Nationalist Party's Central Army Officer Academy and colleges of military specialties), edited by Wen Wen, 2–5. Beijing: Zhongguo wenshi chubanshe, 2010.

Xiao, Jun. *Bayuede xiangcun* (Village in August). Nanjing: Jiangsu wenyi chubanshe, 2010.

———. *Ren yu renjian: Xiao Jun huiyilu* (The person and his life: Memoirs of Xiao Jun). Beijing: Zhongguo wenlian chubanshe, 2006.

Xie, Yingbai. "1929 zhi 1933 nian de Nanjing zhongyang junxiao" (The Central Army Officer Academy at Nanjing between 1929 and 1933) [1963]. In *Guomindang zhongyang lujun xuexiao yu junshi zhuanke xuexiao* (The Nationalist Party's Central Army Officer Academy and colleges of military specialties), edited by Wen Wen, 10–13. Beijing: Zhongguo wenshi chubanshe, 2010.

Xinshenghuo yundong funü zhidaohui. *Rongyu junren fuwu gongzuo jishi* (Records of the work of serving honorable soldiers). Chongqing: N.p., 1944.

Xu, Chonghao. *Zhengbing zhi yange ji shixingfa* (The transformation of the military recruiting system and its implementation). Nanjing: Minzhi shuju, 1929.

Yan'an wenyi congshu bianweihui, ed. *Baogaowenxue* (Documentary literatures). Yan'an wenyi congshu (Series of Yan'an literatures and arts), vol. 6. Changsha: Hunan renmin chubanshe, 1983.

———. *Wenyi lilun* (Theories on literature and arts). Yan'an wenyi congshu (Series of Yan'an literatures and arts), vol. 1. Changsha: Hunan renmin chubanshe, 1983.

———. *Yanggeju* (The *yangge* drama). Yan'an wenyi congshu (Series of Yan'an literatures and arts), vol. 7. Changsha: Hunan renmin chubanshe, 1983.

Ye, Shoukang, et al. *Junren shouce* (Handbooks for soldiers). Jinhua: Zhejiang junxun tushu chubanshe, 1939.

Yi, Qun, ed. *Zhandou de suhui* (Quick sketches of the combat). Chongqing: Zuojia shuwu, 1943.

Yu, Zhaoming. *Rongyu junren jiuye fudao* (Assisting honorable soldiers in getting employed). Nanjing: Zhengzhong shuju, 1947.

——. *Rongyu junren zhi zhiye zaizao* (Vocational rehabilitation for honorable soldiers). N.p.: Junshi weiyuanhui houfang qinwu bu zhengzhi bu, 1942.

Zhao, Chaogou. *Yan'an yiyue* (One month in Yan'an). Washington D.C.: Association of Research Libraries, 1969.

Zhao, Zhen. "Huiyi Jiang Jieshi dui zhongyang junxiao xueshengde longluo shouduan" (Memories on the techniques employed by Chiang Kai-shek to win over the cadets at the Central Army Officer Academy) [1964]. In *Guomindang zhongyang lujun xuexiao yu junshi zhuanke xuexiao* (The Nationalist Party's Central Army Officer Academy and colleges of military specialties), edited by Wen Wen, 12–16. Beijing: Zhongguo wenshi chubanshe, 2010.

Zhongguo di'er lishi dang'an guan, ed. *Zhonghua minguoshi dang'an ziliao huibian* (Compilations of archives and materials during the Republic of China period), vol. 5, no. 1. Nanjing: Jiangsu guji chubanshe, 1994.

Zhongguo Guomindang zhongyang weiyuanhui dangshi weiyuanhui, ed. *Guofu quanji* (Complete compilation of the works by the Father of Republic of China Sun Yat-sen), vol. 2. Taibei: Zhongguo Guomindang zhongyang weiyuanhui dangshi weiyuanhui, 1981.

Zhongguo shehui kexueyuan jindaishisuo, ed. *Sun Yat-sen quanji* (Complete Compilation of Sun Yat-sen), vol. 10. Beijing: Zhonghua shuju, 1986.

Zhongguo xiandai wenxueguan, ed. *Qiu Dongping daibiaozuo* (Representative works by Qiu Dongping). Beijing: Huaxia chubanshe, 2011.

——. *Xie Bingying daibiaozuo* (Representative works by Xie Bingying). Zhongguo xiandai wenxueguan (Beijing: Huaxia chubanshe, 2009.

Zhongyang xunliantuan bingyi ganbu xunlianban, ed. *Bingyi fagui huibian, yiwu* (Compilation of laws and regulations on conscription, drafting affairs). N.p.: N.p., 1942.

Zhu, Peide. *Junguan de xinshenghuo* (New life of military officers). Nanjing: Zhengzhong shuju, 1934.

Chinese-Language Secondary Sources

Chen, Ningsheng. *Jiang Jieshi he Huangpu xi* (Chiang Kai-shek and his Whampoa clique). Zhengzhou: Henan renmin chubanshe, 1994.

Chen, Yuhuan. *Huangpu junxiao diyiqisheng yanjiu* (Studies on the cadets in the first class of the Whampoa Military Academy). Guangzhou: Zhongshan daxue chubanshe, 2007.

Du, Yuanzai. "Kangzhan shiqi zhi qingnian huodong" (The activities of the youth during the Anti-Japanese War). In *Geming wenxian* (Historical documents of the revolution), vol. 62, edited by Luo Jialun, 193–195. Taibei: Zhongyang wenwu gongyingshe, 1973.

Gao, Hua. *Hong taiyang shi zenyang shengqi de: Yan'an zhengfeng yundong de lailong qumai* (How the red sun rose: The origins and development of the Yan'an Rectification Movement). Hong Kong: Zhongwen daxue chubanshe, 2000.

Gou, Xingchao. "Kangzhan shiqide suican bufei yundong" (The campaign of disabled-but-not-useless during the Second Sino-Japanese War). *Wenshi zazhi* (Journal of literature and history) 5 (2007): 53–55.

Guo, Shaoyi. *Qingnian yuanzhengjun zhilüe* (The history of the Educated Youth Expeditionary Army). Taibei: Youshi wenhua shiye gongsi, 1987.

Jiang, Pei. "Zhanshi zhishi qingnian congjun yundong shuping" (A study on the Campaign to Mobilize Educated Youths to Join the Army during the Anti-Japanese War). *Kangri zhanzheng yanjiu* (Studies on the Anti-Japanese War) 1 (2004): 61–95.

Liu, Zhongshu, ed. *Zhongguo xiandai baibu zhongchangpian xiaoshuo lunxi, shang* (Reviews of a hundred novellas and novels in modern China, part 1). Changchun: Jilin daxue chubanshe, 1986.

Peng, Xiuliang. "Duan Shengwu" (Biography of Duan Shengwu). *Tuixiu shenghuo* (Retirement life) 12 (2010): 34–36.

Peng, Xunhou. *Shijie fanfaxisi zhanzheng zhong de Zhongguo* (China in the World Anti-Fascism War). Beijing: Wuzhou chuanbo chubanshe, 2005.

Ren, Zipeng. "Dageming shidai Huangpu nübing" (Whampoa women cadets in the revolutionary era). *Wenhua Zhongguo* (Chinese culture) (July 14, 2012). http://cul.china.com.cn/lishi/2012-07/14/content_5161009.htm. Accessed October 12, 2012.

Shu, Yang, ed. *Huangpu junxiao yanjiu* (Studies on the Whampoa Military Academy), vol. 3. Guangzhou: Zhongshan daxue chubanshe, 2008.

Sun, Yuqin. "Kangzhan moqide shiwan zhishi qingnian congjun yundong shuping" (A review on the Campaign to Mobilize Educated Youths to Join the Army in the last stage of the Anti-Japanese War). *Kangri zhanzheng yanjiu* (Studies on the Anti-Japanese war) 3 (2010): 19–27.

Wang, Ke, and Xu Sai. *Xiao Jun pingzhuan* (Biography of Xiao Jun). Chongqing: Chongqing chubanshe, 1993.

Wang, Xiaogui, and Gong Zeqi. *Zhongguo jindai junren daiyu shi* (History of the treatment of soldiers in modern China). Beijing: Haichao chubanshe, 2006.

Wei, Rulin, ed. *Zhanshi lunji* (Collection of the studies on the history of wars). Taibei: Huagang chuban youxian gongsi, 1976.

Xue, Xiaoxu, and Du Lei. "Yan'an shiqi de xin yangge yundong jiedu" (Interpretation of the new *yangge* movement in Yan'an). *Wenshi ziliao* (Materials on literature and history) 33 (November 2008): 78–79.

Zhao, Xiaoyang. "Kangri zhanzheng shiqi Zhongguo jidujiao qingnianhui junren fuwubu yanjiu" (The study on the Emergency Service to Soldiers Program of the Chinese YMCA during the Anti-Japanese War). *Kangri zhanzheng yanjiu* (Studies on the Anti-Japanese War) 2 (2011): 30–39.

Zhongguo shehui kexueyuan xinwen yanjiusuo zhongguo baokanshi yanjiushi, ed. *Yan'an wencui* (Literature collections in Yan'an). Beijing: Beijing chuban-she, 1984.

English-Language Primary Sources

Belden, Jack. *China Shakes the World*. New York: Monthly Review Press, 1970.

Carlson, Evans Fordyce. *The Chinese Army, Its Organization and Military Efficiency*. New York: Institute of Pacific Relations, 1940.

Chang, Sidney, and Ramon Hawley Myers, eds. *The Storm Clouds Clear over China: The Memoir of Chen Lifu, 1900–1993*. Stanford, Calif.: Hoover Institution Press, 1994.

Chen, Lang Tung. "Christian Education in China." *Chinese Students' Monthly* 18, no. 3 (1923): 20–25.

Johnston, R. F. "The Cult of Military Heroes in China." *New China Review* 3 (February 1921): 57.

White, Theodore H. *Thunder out of China*. New York: Da Capo Press, 1980.

Xiao, Jun. *Village in August*. Translated by Tien Chün. New York: Smith and Durrell, 1942.

Xie, Bingying. *A Woman Soldier's Own Story*. Translated by Lily Chia Brissman and Barry Brissman. New York: Columbia Univ. Press, 2001.

English-Language Secondary Sources

Ashe, Fidelma, *The New Politics of Masculinity: Men, Power and Resistance*. New York: Routledge, 2007.

Bian, Morris. "Building State Structure: Guomindang Institutional Rationalization during the Sino-Japanese War, 1937–1945." *Modern China* 31, no. 1 (January 2005): 35–71.

——. *The Making of the State Enterprise System in Modern China: The Dynamics of Institutional Change*. Cambridge, Mass.: Harvard Univ. Press, 2005.

Bichler, Lorenz. "Coming to Terms with a Term: Notes on the History of the Use of Socialist Realism in China." In *In the Party Spirit: Socialist Realism and Literary Practice in the Soviet Union, East Germany and China,* edited by Hilary Chung, Michael Falchikov, B. S. McDougall, and K. McPherson, 30–43. Amsterdam: Rodopi B. V., 1996.

Bieler, Stacey. *"Patriots" or "Traitors"?: A History of American-Educated Chinese Students*. New York: Routledge, 2009.

Bonavia, David. *China's Warlords*. Hong Kong: Oxford Univ. Press, 1995.

Chamberlain, Heath. "Civil Society with Chinese Characteristics." *China Journal* 39 (January 1998): 69–81.

Chan, Ming K., and Arif Dirlik. *Schools into Fields and Factories: Anarchists, the Guomindang, and the National Labor University in Shanghai, 1927–1932*. Durham, N.C.: Duke Univ. Press, 1991.

Chang, Jui-te. "Bombs Don't Discriminate? Class, Gender, and Ethnicity in the Air-Raid Shelter Experiences of the Wartime Chongqing Population." In *Beyond Suffering: Recounting War in Modern China,* edited by James Flath and Norman Smith, 59–79. Vancouver, Canada: Univ. of British Columbia Press, 2011.

———. "Nationalist Army Officers during the Second Sino-Japanese War, 1937–1945." *Modern Asian Studies* 30, no. 4 (October 1996): 1033–1056.

Chen, Yung-fa. *Making Revolution: The Communist Movement in Eastern and Central China, 1937–1945.* Berkeley: Univ. of California Press, 1986.

Cheng, Yinghong. *Creating the "New Man": From Enlightenment Ideals to Socialist Realities.* Honolulu: Univ. of Hawai'i Press, 2009.

Chi, Hsi-sheng. *Warlord Politics in China, 1916–1928.* Stanford, Calif.: Stanford Univ. Press, 1976.

Chien, Tuan-sheng. *The Government and Politics of China, 1912–1949.* Stanford, Calif.: Stanford Univ. Press, 1950.

Chung, Jae Ho, and Tao-chiu Lam, eds. *China's Local Administration: Traditions and Changes in the Sub-National.* New York: Routledge, 2010.

Coble, Parks. "Chiang Kai-shek and the Anti-Japanese Movement in China: Zou Taofen and the National Salvation Association, 1931–1937." *Journal of Asian Studies* 44, no. 2 (February 1985): 293–310.

Cong, Xiaoping. *Teachers' Schools and the Making of the Modern Chinese Nation-State, 1897–1937.* Vancouver, Canada: Univ. of British Columbia Press, 2007.

Diamant, Neil. *Embattled Glory: Veterans, Military Families, and the Politics of Patriotism in China, 1949–2007.* Lanham, Md.: Rowman and Littlefield, 2009.

Dillon, Michael. *China: A Modern History.* New York: I. B. Tauris, 2010.

Duara, Prasenjit. *Culture, Power and the State: Rural North China, 1900–1942.* Stanford, Calif.: Stanford Univ. Press, 1988.

———. *Rescuing History from the Nation: Questioning Narratives of Modern China.* Chicago: Univ. of Chicago Press, 1995.

Eastman, Lloyd. *Seeds of Destruction: Nationalist China in War and Revolution, 1937–1949.* Stanford, Calif.: Stanford Univ. Press, 1984.

Eastman, Lloyd, et al., eds. *The Nationalist Era in China, 1927–1949.* New York: Cambridge Univ. Press, 1991.

Ebrey, Patricia. *Confucianism and Family Rituals in Imperial China: A Social History of Writing about Rites.* Princeton, N.J.: Princeton Univ. Press, 1991.

Esherick, Joseph. "Ten Theses on the Chinese Revolution." *Modern China* 21, no. 1 (January 1995): 45–76.

Fairbank, John K. *The United States and China.* Cambridge, Mass.: Harvard Univ. Press, 1983.

Flath, James, and Norman Smith, eds. *Beyond Suffering: Recounting War in Modern China.* Vancouver, Canada: Univ. of British Columbia Press, 2011.

Ford, Daniel. *Flying Tigers: Claire Chennault and the American Volunteer Group.* Washington, D.C.: Smithsonian Institution Press, 1991.

Gao, James Zheng. *Historical Dictionary of Modern China (1800–1949)*. Lanham, Md.: Scarecrow Press, 2009.

Gillin, Donald. *Warlord: Yen Hsi-shan in Shansi Province, 1991–1949*. Princeton, N.J.: Princeton Univ. Press, 1970.

Gittings, John. *The Role of the Chinese Army*. London: Oxford Univ. Press, 1967.

Glosser, Susan. *Chinese Visions of Family and State, 1915–1953*. Berkeley: Univ. of California Press, 2003.

Goldman, Merle, and Leo Ou-fan Lee, eds. *An Intellectual History of Modern China*. New York: Cambridge Univ. Press, 2002.

Goldstein, Joshua. *War and Gender: How Gender Shapes the War System and Vice Versa*. New York: Cambridge Univ. Press, 2001.

Goodman, Bryna. "Creating Civic Ground: Public Maneuverings and the State in the Nanjing Decade." In *Remapping China: Fissures in Historical Terrain*, edited by Gail Hershatter et al., 164–180. Stanford, Calif.: Stanford Univ. Press, 1996.

———. *Native Place, City, and Nation: Regional Networks and Identities in Shanghai, 1853–1937*. Berkeley: Univ. of California Press, 1995.

Graff, David, and Robin Higham, eds. *A Military History of China*. Lexington: Univ. Press of Kentucky, 2012.

Hanyu, Mitsutoshi. "From Campus to Battle: Student Mobilization and Transition of Japanese Imperial Military's Draft Policies." *Military History Research Annual Report*, vol. 12 (March 2009). http://www.nids.mod.go.jp/publication/senshi/pdf/200903/08.pdf. Accessed September 21, 2016.

Hammond, Kenneth, and Kristin Stapleton, eds. *The Human Tradition in Modern China*. Lanham, Md.: Rowman and Littlefield, 2007.

Holm, David. *Art and Ideology in Revolutionary China*. Oxford: Clarendon Press, 1991.

Hsia, Chih-tsing. *A History of Modern Chinese Fiction*. Bloomington: Indiana Univ. Press, 1999.

Hsiung, James. "The War and After: World Politics in Historical Perspective." In *China's Bitter Victory: The War with Japan, 1937–1941*, edited by James Hsiung and Steven Levine, 295–306. Armonk, N.Y.: M. E. Sharpe, 1992.

Huebner, Andrew. *The Warrior Image: Soldiers in American Culture from the Second World War to the Vietnam Era*. Chapel Hill: Univ. of North Carolina Press, 2008.

Hung, Chang-tai. *War and Popular Culture: Resistance in Modern China, 1937–1945*. Berkeley: Univ. of California Press, 1994.

Ip, Hung-Yok. *Intellectuals in Revolutionary China: Leaders, Heroes and Sophisticates*. London: Routledge, 2005.

Johnson, Chalmers. *Peasant Nationalism and Communist Power: The Emergence of Revolutionary China*. Stanford, Calif.: Stanford Univ. Press, 1962.

Jowett, Philip. *The Chinese Army, 1937–1949: World War II and Civil War*. Oxford: Osprey Publishing, 2005.

——. *Chinese Civil War Armies, 1911–1949.* Oxford: Osprey Publishing, 1997.

——. *Chinese Warlord Armies, 1911–1930.* Oxford: Osprey Publishing, 2010.

Kaske, Elisabeth. *The Politics of Language in Chinese Education, 1895–1919* (Leiden, Netherlands: Koninklijke Brill NV, 2008), 142.

Kimmel, Michael, and Michael Messner, eds. *Men's Lives.* Boston: Allyn and Bacon, 1995.

Kirby, William. *Germany and Republican China.* Stanford, Calif.: Stanford Univ. Press, 1984.

Lary, Diana. *China's Civil War: A Social History, 1945–1949.* Cambridge, UK: Cambridge Univ. Press, 2015.

——. *Region and Nation: The Kwangsi Clique in Chinese Politics, 1925–1937.* Cambridge, UK: Cambridge Univ. Press, 1974.

——. *Warlord Soldiers: Chinese Common Soldiers, 1911–1937.* Cambridge, UK: Cambridge Univ. Press, 1985.

Laughlin, Charles A. *Chinese Reportage: The Aesthetics of Historical Experience.* Durham, N.C.: Duke Univ. Press, 2002.

Lean, Eugenia. *Public Passions: The Trial of Shi Jianqiao and the Rise of Popular Sympathy in Republican China.* Berkeley: Univ. of California Press, 2007.

Leary, William. *The Dragon's Wing: The China National Aviation Corporation and Development of Commercial Aviation in China.* Athens: Univ. of Georgia Press, 1976.

Levich, Eugene. *The Kwangsi Way in Kuomintang China, 1931–1939.* Armonk, N.Y.: M. E. Sharpe, 1993.

Li, Lincoln. *Student Nationalism in China, 1924–1949.* Albany: State Univ. of New York Press, 1994.

Li, Xiaobing. *A History of the Modern Chinese Army.* Lexington: Univ. Press of Kentucky, 2007.

Liu, Frederick Fu. *A Military History of Modern China.* Princeton, N.J.: Princeton Univ. Press, 1956.

Liu, Yu. "Maoist Discourse and the Mobilization of Emotions in Revolutionary China." *Modern China* 36, no. 3 (May 2010): 329–362.

Louie, Kam. *Theorising Chinese Masculinity: Society and Gender in China.* Cambridge, UK: Cambridge Univ. Press, 2002.

MacKinnon, Stephen. "Refugee Flight at the Outset of the Anti-Japanese War." In *The Scars of War: The Impact of Warfare on Modern China,* edited by Diana Lary and Stephen MacKinnon, 118–135. Vancouver, Canada: Univ. of British Columbia Press, 2002.

——. *Wuhan, 1938: War, Refugees, and the Making of Modern China.* Berkeley: Univ. of California Press, 2008.

Mann, Susan. "The Male Bond in Chinese History and Culture." *American Historical Review* 105, no. 5 (December 2000): 1604.

McCormack, Gavan. *Chang Tso-lin in Northeast China, 1911–1928.* Folkestone, UK: Dawson and Sons, 1977.

Mulready-Stone, Kristen. *Mobilizing Shanghai Youth: CCP Internationalism, GMD Nationalism and Japanese Collaboration*. New York: Routledge, 2015.

Pepper, Suzanne. *Civil War in China: The Political Struggle, 1945–1949*. 2nd ed. Lanham, Md.: Rowman and Littlefield, 1999.

Perry, Elizabeth. "Moving the Masses: Emotion Work in the Chinese Revolution." *Mobilization: An International Quarterly* 7, no. 2 (summer 2002): 111–128.

Pletcher, Kenneth, ed. *The History of China*. New York: Britannica Educational Publishing, 2011.

Potter, Sulamith Heins. "Cultural Construction of Emotion in Rural Chinese Social Life." *Ethos* 16, no. 2 (June 1988): 208.

Rigdon, Susan. "National Salvation: Teaching Civic Duty in China's Christian Colleges." In *China's Christian Colleges: Cross-Cultural Connections, 1900–1950*, edited by Daniel Bays and Ellen Widmer, 193–217. Stanford, Calif.: Stanford Univ. Press, 2009.

Rowe, William. "Public Sphere in Modern China." *Modern China* 16, no. 3 (July 1990): 309–329.

Rummel, R. J. *Death by Government*. New Brunswick, N.J.: Transaction Publishers, 2009.

Schumpeter, Joseph. *Capitalism, Socialism, and Democracy*. New York: Harper and Row, 1975.

Schwartz, Benjamin. *Chinese Communism and the Rise of Mao*. Cambridge, Mass.: Harvard Univ. Press, 1951.

Scobell, Andrew. *China's Use of Military Force: Beyond the Great Wall and the Long March*. Cambridge, UK: Cambridge Univ. Press, 2003.

Sebesta, Seungsook Moon. *Militarized Modernity and Gendered Citizenship in South Korea*. Durham, N.C.: Duke Univ. Press, 2005.

Selden, Mark. *China in Revolution: The Yenan Way Revisited*. New York: M. E. Sharpe, 1995.

Sheridan, James. *Chinese Warlord: The Career of Feng Yu-hsiang*. Stanford, Calif.: Stanford Univ. Press, 1966.

Shu, Yunzhong. *Buglers on the Home Front: The Wartime Practice of the Qiyue School*. Albany: State Univ. of New York Press, 2000.

Sieber, Patricia, ed. *Red Is Not the Only Color: Contemporary Chinese Fiction on Love and Sex between Women, Collected Stories*. Lanham, Md.: Rowman and Littlefield, 2001.

Snow, Edgar. "Introduction." In Xiao Jun, *Village in August*. Translated by Tien Chün. New York: Smith and Durrell, 1942.

Stranahan, Patricia. *Underground: The Shanghai Communist Party and the Politics of Survival, 1927–1937*. Lanham, Md.: Rowman and Littlefield, 1998.

Strand, David. "'A High Place Is No Better Than a Low Place': The City in the Making of Modern China." In *Becoming Chinese: Passages to Modernity and Beyond*, edited by Wen-hsin Yeh, 98–136. Berkeley: Univ. of California Press, 2000.

Strauss, Julia. *Strong Institutions in Weak Polities: State Building in Republican China, 1927–1940.* Oxford: Clarendon Press, 1998.

Sutton, Donald. *Provincial Militarism and the Chinese Republic: The Yunnan Army.* Ann Arbor: Univ. of Michigan Press, 1980.

Taylor, Jay. *The Generalissimo's Son: Chiang Ching-kuo and the Revolutions in China and Taiwan.* Cambridge, Mass.: Harvard Univ. Press, 2000.

Van de Ven, Hans. "The Military in the Republic." *China Quarterly* 150 (June 1997): 352–374.

———. *War and Nationalism in China, 1925–1945.* New York: Routledge Curzon, 2003.

———, ed. *Warfare in Chinese History.* Leiden, Netherlands: Koninklijke Brill NV, 2000.

Wakeman, Frederic. *Spymaster: Dai Li and the Chinese Secret Service.* Berkeley: Univ. of California Press, 2003.

Wakin, Malham, ed. *War, Morality, and the Military Professor.* Boulder, Colo.: Westview Press, 1986.

Waldron, Arthur. *From War to Nationalism: China's Turning Point.* Cambridge, UK: Cambridge Univ. Press, 1995.

———. "The Warlord: Twentieth-Century Chinese Understanding of Violence, Militarism, and Imperialism." *American Historical Review* 96, no. 4 (1991): 1073–1100.

Wang, Jing M. *When "I" Was Born: Women's Autobiography in Modern China.* Madison: Univ. of Wisconsin Press, 2008.

Wang, Ke-wen. *Modern China: An Encyclopedia of History, Culture, and Nationalism.* New York: Garland, 1998.

Wang, Lingzhen. *Personal Matters: Women's Autobiographical Practice in Twentieth-Century China.* Stanford, Calif.: Stanford Univ. Press, 2004.

Washburn, Dennis, and A. Kevin Reinhart, eds. *Converting Cultures: Religion, Ideology, and Transformations of Modernity.* Leiden, Netherlands: Koninklijke Brill NV, 2007.

Wasserstrom, Jeffrey, and Susan Brownell, eds. *Chinese Femininities, Chinese Masculinities: A Reader.* Los Angeles: Univ. of California Press, 2002.

Wou, Odoric Y. K. *Militarism in Modern China: The Career of Wu Pei-fu.* Folkestone, UK: Dawson and Sons, 1978.

Xiao, Yanzhong. "Recent Mao Zedong Scholarship in China." In *A Critical Introduction to Mao,* edited by Timothy Creek, 273–287. New York: Cambridge Univ. Press, 2000.

Xu, Guangqiu. *War Wings: The United States and Chinese Military Aviation, 1929–1949.* Westport, Conn.: Greenwood Press, 2001.

Xu, Xiaoqun. *Chinese Professionals and the Republican State: The Rise of Professional Associations in Shanghai, 1912–1937.* New York: Cambridge Univ. Press, 2004.

Yeh, Wen-hsin. "Writing in Wartime China: Chongqing, Shanghai, and Southern

Zhejiang." Paper presented at the CCKF (Chiang Ching-kuo Foundation for International Scholarly Exchange)-CHCI (Consortium of Humanities Centers and Institutes) Summer Institute, *China in a Global World War II,* Cambridge, UK, July 2017.

Young, Arthur. *China's Nation Building Effort, 1927–1937: The Financial and Economic Papers.* Stanford, Calif.: Hoover Institution Press, 1971.

Zarrow, Peter. *China in War and Revolution, 1895–1949.* New York: Routledge, 2005.

Index

ASIA IN THE NEW MILLENNIUM

SERIES EDITOR: Shiping Hua, University of Louisville

Asia in the New Millennium is a series of books offering new interpretations of an important geopolitical region. The series examines the challenges and opportunities of Asia from the perspectives of politics, economics, and cultural-historical traditions, highlighting the impact of Asian developments on the world. Of particular interest are books on the history and prospect of the democratization process in Asia. The series also includes policy-oriented works that can be used as teaching materials at the undergraduate and graduate levels. Innovative manuscript proposals at any stage are welcome.

ADVISORY BOARD
William Callahan, University of Manchester, Southeast Asia and Thailand
Lowell Dittmer, University of California at Berkeley, East Asia and South Asia
Robert Hathaway, Woodrow Wilson International Center for Scholars, South Asia, India, and Pakistan
Mike Mochizuki, George Washington University, East Asia, Japan, and Korea
Peter Moody, University of Notre Dame, China and Japan
Brantly Womack, University of Virginia, China and Vietnam
Charles Ziegler, University of Louisville, Central Asia and Russia Far East

BOOKS IN THE SERIES

The Future of China-Russia Relations
Edited by James Bellacqua

North Korea and the World: Human Rights, Arms Control, and Strategies for Negotiation
Walter C. Clemens Jr.

Contemporary Chinese Political Thought: Debates and Perspectives
Edited by Fred Dallmayr and Zhao Tingyang

Power versus Law in Modern China: Cities, Courts, and the Communist Party
Qiang Fang and Xiaobing Li

China Looks at the West: Identity, Global Ambitions, and the Future of Sino-American Relations
Christopher A. Ford

The Mind of Empire: China's History and Modern Foreign Relations
Christopher A. Ford

State Violence in East Asia
Edited by N. Ganesan and Sung Chull Kim

Challenges to Chinese Foreign Policy: Diplomacy, Globalization, and the Next World Power
Edited by Yufan Hao, C. X. George Wei, and Lowell Dittmer

The Price of China's Economic Development: Power, Capital, and the Poverty of Rights
Zhaohui Hong

Japan after 3/11: Global Perspectives on the Earthquake, Tsunami, and Fukushima Meltdown
Edited by Pradyumna P. Karan and Unryu Suganuma

Korean Democracy in Transition: A Rational Blueprint for Developing Societies
HeeMin Kim

Modern Chinese Legal Reform: New Perspectives
Edited by Xiaobing Li and Qiang Fang

Democracy in Central Asia: Competing Perspectives and Alternative Strategies
Mariya Y. Omelicheva

China's Encounter with Global Hollywood: Cultural Policy and the Film Industry, 1994–2013
Wendy Su

Growing Democracy in Japan: The Parliamentary Cabinet System since 1868
Brian Woodall

The Soldier Image and State-Building in Modern China, 1924–1945
Yan Xu

Inside China's Grand Strategy: The Perspective from the People's Republic
Ye Zicheng, Edited and Translated by Steven I. Levine and Guoli Liu

Civil Society and Politics in Central Asia
Edited by Charles E. Ziegler